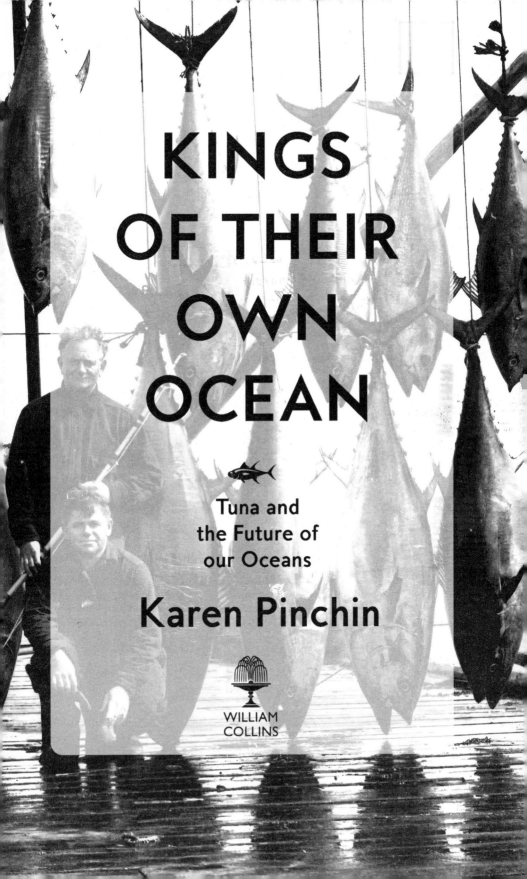

KINGS
OF THEIR
OWN
OCEAN

Tuna and
the Future of
our Oceans

Karen Pinchin

WILLIAM
COLLINS

William Collins
An imprint of HarperCollins*Publishers*
1 London Bridge Street
London SE1 9GF

WilliamCollinsBooks.com

HarperCollins*Publishers*
Macken House
39/40 Mayor Street Upper
Dublin 1
D01 C9W8
Ireland

First published in Great Britain in 2023 by William Collins
First published in the United States in 2023 by Dutton,
an imprint of Penguin Random House LLC

1

Neruda, Pablo. "Oda a un gran atún en el mercado." *Tercer libro de las odas.* © Pablo
Neruda, 1957, and Fundación Pablo Neruda. (Translation by Robin Robertson.)

Eyüboğlu, Bedri Rahmi. "The Saga of Istanbul." *A Brave New Quest: 100
Modern Turkish Poems,* edited and translated by Talat S. Halman, associate
editor Jayne L. Warner, 33-39. Syracuse University Press, 2006. © Syracuse
University Press. Reproduced with permission from the publisher.

A catalogue record for this book is available from the British Library

ISBN 978-0-00-846781-4

Book Design by Lorie Pagnozzi

Printed and Bound in the UK using 100% Renewable Electricity
at CPI Group (UK) Ltd

This book is produced from independently certified FSC™ paper
to ensure responsible forest management.

For more information visit: www.harpercollins.co.uk/green

For my father, Don Pinchin,

a

star

on

the

sea

Dead

in front of me,

catafalqued king

of my own ocean;

. . . .

a well-oiled ship of the wind,

the only

true

machine

of the sea: unflawed,

undefiled,

navigating now

the waters of death.

**—PABLO NERUDA, "ODE TO A LARGE TUNA
IN THE MARKET"**

CONTENTS

CONTENTS

AUTHOR'S NOTE

In this book, I use the terms "fisherman" and "fishermen" to refer to both men and women engaged in the act of catching fish, despite the fact that those labels largely obscure the latter's massive importance within global fisheries. (Women compose half the world's total working population in the seafood industry.) Throughout my reporting, many women I interviewed expressed preference for the moniker; I have also used the gender-neutral "angler" when appropriate.

For clarity, I've used imperial pounds for fish prices and weights, reflecting current industry practice in many parts of the world. Otherwise I've used metric units—tonnes, for instance—throughout.

AN ENDING

~~~~~~~~~~

In the middle of the Atlantic Ocean, between swells of blue-gray water, the sky is a dome over a dinner plate, vast and variable. The horizon's flatness cleaves a line through the elements, a trick of perspective obscuring the earth's true curvature. On an unpredictable sea, the permanence of that line, its physical declaration of up and down, anchors a reality a seafarer can cling to, a reality without which she—or you, or I—could lose herself.

At exactly 45° west longitude, between the coasts of North America and Europe and hundreds of kilometers from the nearest land, there's another line. This one cuts the ocean perpendicular to the horizon and runs vertically from surface to seafloor. Plummeting downward along this vertical cross section, dropping through watery layers, white sprays are tossed from wave tops; clouds of plankton and krill drift languidly; the sunlight fades and eventually disappears. Spidery starfish creep along the sandy seafloor past lurking wolffish. From spitting surface to deepest undersea canyon, these strata take shape in my imagination like a painting in one of my son's picture books.

Ever since this line at the 45° west meridian was invented by humans, it has dictated the fortunes and fate of Atlantic bluefin tuna, one of our planet's most spectacular and vital apex predators. The line is also entirely theoretical, except to the extent that it controls a billion-dollar industry and has allowed humans to run roughshod over one of the world's most contested species for decades.

Atlantic bluefin tuna, *Thunnus thynnus,* have lived for millions of years in our physical world and captured human imaginations long before the written word. Ancient peoples trapped bluefin in shallows and killed them with rocks and sticks; Greeks and Phoenicians immortalized the economically important fish in art and theater, and even stamped its image on early currency. Bluefin are "charismatic megafauna"—a category of large, big-eyed beasts including elephants, lions, and whales that serve as placeholders in the public imagination—and have come to represent the health of our oceans and the overfishing that besieges them.

Compared with floppy, schooling fish, giant bluefin are more machine than fuel, more predator than prey. I was in my mid-20s when I first saw one in person, and it fit my idea of "fish" as much as a dinosaur resembles a chickadee; there was, from that initial impression, a problem of scale. The largest tuna ever caught weighed 1,496 pounds, if you can imagine a grand piano shaped like a nuclear weapon. To stand beside a just-landed giant bluefin, still slick from salt water, feels akin to standing beside a natural marvel like Niagara Falls or an erupting volcano. There's beauty, but also danger.

Every late summer and early fall, schools of gorgeous, massive bluefin rocket along North America's Atlantic coast. They are an advanced rod-and-reel sport fisherman's fish: catching one requires time and a full set of pricey and intensely specific gear, including a volleyball-sized brass reel built to handle a tuna's heft. Some fishermen call bluefin "marauders"; others, "racers." The genus name *Thunnus*, after all, comes from the ancient Greek word θύνω, or *thuno*: rush.

Throughout the mid and late 1970s, the world's three bluefin tuna populations—Atlantic, Pacific, and Southern—plunged globally because of rampant overfishing, largely due to demand for the sushi-grade fish in Japan. In 1981, two dozen tuna-fishing countries made their first move to restrict the trade, drawing a line down the middle of the Atlantic Ocean during a meeting hosted by the Spanish government in Madrid. That line ran from the southernmost tip of Greenland, zigzagged around Brazil's rocky bulge, then progressed south off the map toward Antarctica. Starting the following year, all bluefin tuna caught to the right of the line toward Europe and North Africa would be considered "eastern" fish, while all to the left would be "western." At the time, prevailing scientific thought held that the powerful, warm-blooded fish largely gravitated back to the coastlines where they were born, with eastern fish spawning in the balmy Mediterranean and their western counterparts in the Gulf of Mexico.

Overnight, this conceptual line transformed the fisheries for Atlantic bluefin tuna. Toward the European side of the Atlantic Ocean, harvests remained a free-for-all with no quotas on how many fish could be caught and killed for the countries that had historically fished those waters. Toward North America, for countries including the United States and Canada—and for historical reasons, Japan—not a single fish, other than those purportedly caught for "scientific purposes," could be legally caught. Called the two-stock theory, this understanding of how tuna breed and migrate, as two almost entirely separate populations, set the stage for how Atlantic bluefin have been protected and managed for the entirety of my lifetime.

Growing up in a leafy suburb abutting Lake Ontario, I learned how to catch and gut a fish before I knew how to blow-dry my hair or paint my nails. My father, Don, a successful businessman and scientist, had spent his own childhood angling on rural rivers and lakes with his farmer parents and wanted to pass along his love of fishing to me and my two younger siblings. We spent summers jamming oozing worms onto hooks, threading bobbers, and paddling to find the best fishing holes. Fishing, he lectured, was a controllable craft and act of human will, hypothesis testing, and logic—with results you could eat for supper. In this and other ways, raising us alongside my chemist mother, he tried to teach us about the immutable final authority of science. In contrast, I wanted to write poetry or maybe books, captivated by our world's living systems and its nuanced, pulsing rhythms. My parents rolled their eyes at my earnestness.

Still, I refused to give up my dream of being a writer. Curious about food, I enrolled in culinary school and used that training to write about chefs and restaurants while earning extra money cooking and catering. Once, a fisherman supplied a three-foot-long bluefin tuna to a restaurant where I worked, forcing me to search online for an instructional video on how to butcher the pricey fish. Before inserting my razor-sharp knife along its backbone, I stroked its metallic, glossy sides, marveling at its heft and beauty. With dense, red flesh, it was nothing like the flimsy perch and bass I had grown up catching. Once I had cut and vacuum-sealed the bluefin's slabbed steaks, they all sold out within hours. I scraped extra flesh from the carcass with a spoon and used it to make a marinated Hawaiian poke. I heaped the unctuous result onto tortilla chips, the richness of its flavor mingling with salty soy sauce and fiery chilies. I felt thankful, and lucky.

In my early 30s, in the months before I gave birth to my son, my father was diagnosed with pancreatic cancer. I found out he was dying on an unexpected video call from my family, who had gathered to break the news. It sucked me under, swamping me with the cosmic unfairness of receiving new life just as the universe demanded one in return. Cancer doesn't run in our family; my father didn't smoke or drink heavily. There seemed to be no logic to it, no equations or science that could explain the loss we faced, even as a new child grew inside me.

Throughout his career, my father had worked to prove himself to the world, to his own small-town farmer parents, to himself, to us. He recited ancient poetry from dining room chairs—his specialty was Lord Byron's "The Destruction of Sennacherib"—collected antique maps and clocks, tore branches loaded with cherries from abandoned orchards and passed them around our car as he drove. He was always blunt and sometimes cruel. He could be a hard person to love.

Pancreatic cancer left my formerly vigorous father jaundiced and irritable, but it was an aggressive glioblastoma brain tumor that eventually hollowed him out. The last time I saw my father was on a sweltering summer day in the city of my birth, my newborn son in my arms. My father died before I had the chance to understand him, the loads he carried, or the invisible energies that spurred him forward. His absence left me unmoored, setting my sense of home and belonging adrift.

Two years after that hot, sad summer, I started a master's program studying science journalism in New York City. My husband, son, and I moved to the Upper West Side on an August day laced with the smell of linden blossoms and melting pavement. Within academia I wrote about the ocean and climate change, about the death and survival of salt-inundated coastal trees, about scientists who studied the North Sea's smallest currents. Recollections of my parents' professional lives that I had walled off from my writing life up to that point came flooding into my journalism: playing with windup toys on the floor of my father's office while he worked on a Saturday; my mother performing experiments and floating in zero gravity on flights at the Johnson Space Center.

My parents allowed their careers to seep into our lives in small ways, like my mother's uncanny skill at tempering chocolate and my father's ability to toss a fishing line into a riverside pool in the exact spot where he knew bass would be resting. He skillfully steered our canoe, holding monofilament line between his lips, waiting for a telltale tug of a biting fish. He thought nothing of pushing a barbed hook through the lips of leopard frogs he'd catch with cupped hands on the shoreline. He kept those frogs alive on the end of his hook so their kicking legs attracted the biggest muskellunges and pikes. We ate those fish for supper after he gutted them on the dock. We poked through their slimy stomach contents to make sure we were using the right bait for the season.

ONE SUNDAY AFTERNOON while walking in Manhattan with my husband and toddler, my cell phone rang with an out-of-the-blue call from a marine scientist named Molly Lutcavage. I had emailed her the week before, curious about her work tagging bluefin tuna off the northeastern coast of the United States near the historic port of Gloucester, Massachusetts. We had never talked before, but she jumped into the call as if we were longtime collaborators. She talked quickly; she told me a story about a blue-fin tuna, a scientific tag, and an astonishing discovery. Half an hour later, as my husband fed my son pizza inside a streetside café, I was sitting there on the sidewalk, my back leaning against the café's cold stone outer wall. With my head crooked to the side, I pressed my slippery phone to my ear above an aching shoulder as I scribbled into a notebook.

Eleven years prior, Lutcavage had been conducting research off the coast of Cape Cod aboard a chartered tuna fishing boat when her team caught an approximately five-year-old Atlantic bluefin tuna. The female fish's iridescent gray-blue skin gleamed and re-fracted the light like a scarab. She fought the air, kicking the small, powerful muscles of her tail in bursts as Lutcavage's team scrambled to keep her still and unharmed. As Lutcavage's boat bobbed on the waves, the scientist realized they had captured an incredible needle in a haystack: a large tuna that had already been tagged once before. In her back nestled a yellow spaghetti tag, a fish-tracking apparatus named for its long, cylindrical shape. Embedded near her dorsal fin, the palm's-length tag bore a string of tiny identifying numbers. For Lutcavage, finding the tag felt "like Christmas," for reasons that wouldn't become obvious to me for months.

From the moment she saw the tag, Lutcavage harbored suspicions about who had inserted it. Back on land, she confirmed those suspicions, even going so far as to personally call the fisherman responsible for tagging the fish three years prior: Al Anderson. His was a name she knew well. Anderson was an industry-famous skipper from Narragansett, Rhode Island, whose main business involved taking charter clients out on his boat, the *Prowler*, and catching lots of fish. Unlike on most charters, though, Anderson's clients didn't keep most of the fish they caught. Instead, they helped Anderson tag them and set them free, pitching them head-first back into the open ocean. The other fishermen in his harbor thought it was a bizarre business model, but over the decades, Anderson had built a reputation doing it. He had even won international awards for tagging more bluefin tuna than any individual fisherman in the world ever had—or would again.

It had been more than a decade since Molly Lutcavage slipped that recaptured tuna back into the ocean off the coast of Massachusetts. And she called me that day in 2018 in response to my casual email, bearing news of her most recent scientific success—that the same tuna had, only a month before, been captured a third and final time in the south of Portugal. That meant Anderson and Lutcavage's fish had traversed the entire Atlantic, defying conventional wisdom and entering the growing ranks of bluefin messing up that orderly, imaginary 1981 line bisecting the ocean into eastern and western zones. She may even have crossed the Atlantic more than once. Lutcavage eventually gave the fish a name: Amelia, for Amelia Earhart, an iconic woman who had crossed the same ocean on currents of a different kind.

From that day, Amelia's crossing, its implications, and Al Anderson consumed my spare time. The gossamer threads linking him, a fisherman obsessed with fish tagging, to a huge dead fish in Europe gradually entrapped my imagination. I traveled to Fords, New Jersey, Al's birthplace. I spent hours in the Rhode Island basement where he kept his records and spooled his own fishing line. I listened to stories about Al—of the demons that chased him, of the regrets he carried and weaponized in the service of his own success—and was reminded of my father's rare but stormy outbursts. I crossed the Atlantic Ocean, hunting for the exact patch of sea where Amelia had been killed, following her path toward the world's second-largest fish market in Madrid, as well as to Tokyo, where many thousands more of her fellow bluefin end up on the chopping block every year. This story drew me in, spun me around, and spat me out, encapsulating the incomprehensible scale and small cruelties of our modern existence on this planet.

SINCE THE 1970S, global bluefin tuna populations have surfed the curling edge of collapse. Bluefin tuna flesh is tasty and nutritious, qualities that helped fuel the rise of the Phoenician and Roman empires. Despite making up only 1 percent of the world's tuna catch—other species include albacore, yellowfin, and bigeye—bluefin tuna makes up nearly two-thirds of its value. Since Mediterranean fishermen employed huge nets to catch bluefin en masse 3,000 years ago, demand for the fish has existed in uneasy balance with its natural ability to reproduce and thrive. And despite billions of dollars spent, decades of scientific research, and numerous campaigns to "Save the Bluefin," that remains as true today as it was decades ago.

Amelia inspired my own journey into the purpose and meaning of science and its limitations, including an exploration of how far we've come—and how far we've yet to go. And as I lost myself in the ocean, traveling alongside Anderson and Lutcavage's fish, I passed signals and signs of greater upheavals: conflict over global fishing rights, bluefin poaching, and the future of our food supply in a changing climate. For years, Amelia crossed the Atlantic Ocean and evaded capture. She ate and spawned and grew. Humans caught her, and then caught her again. Finally they killed her.

Over those same decades, the people who touched her flanks experienced their own triumphs and tragedies, losses and redemptions. If everything my parents taught me about science is true—if facts are indeed the sun around which we all orbit—then puny human emotions can't hold a candle to the ocean's scale and its untouchable power. Yet around every bend of this journey, I found both human and animal existences amplified: the exquisite ache of failure; the tightrope walk of existence; serene joy in our planet's offerings and deliverance; and ultimately how quickly it all can end. What started as a simple fish story showed me a world where bold scientific, commercial, and personal endeavors can crystallize into a single burning vision. But occasionally those visions collide, sometimes leaving their bearers forlorn, other times carrying them to the greatest heights. And often science itself, the tool and technique through which we grasp for certainty in an uncertain world, has little to do with who, or what, wins or loses.

Diving into the history of bluefin and the human communities that have lived alongside the species for millennia, I glimpsed a vision of a reciprocal future, one in which human passions could dovetail with our long-term survival on this planet—a quest that must include the right of wildlife to thrive alongside us. I discovered we all need our own versions of that 45° line: a boundary to anchor us, others to hold us, or even a small sense that this all might mean something in the end.

But before reaching for dry land, we must first enter the sea.

CHAPTER ONE

HOOKED

~~~~~~~~

Al and Amelia, Their Early Years

When you finally see what goes on underwater, you realize
that you've been missing the whole point of the ocean.
—DAVE BARRY, "BLUB STORY"

Deep below the surface of Rhode Island Sound on September 27, 2004, a bristling school of Atlantic bluefin tuna sliced above an inky-blue landscape of sandy sea bottom and glacier-sheared boulders. Brothers swam beside sisters, alongside cousins and distant cousins. They were all only a year or two old, but in the depths off Block Island they were already feared. The warm-blooded species has a voracious appetite, and the juvenile fish ate nearly everything they came across—shrimp, deep-water squid, jellyfish—near constantly, and in constant motion, since bluefin must swim to breathe. Their eyes, the sharpest of all the bony fishes', perceived filtered light from the surface as it dimmed and brightened around them, each night and day like those before.

Within the school, one half-meter-long female fish coasted, her pectoral fins splayed like airplane wings that helped her glide and tweak the power generated by her sickled tail. She had small, chartreuse-yellow triangular points running along the top and bottom of her back and belly in matching rows of prehistoric finlets. Her torpedo-shaped head was smooth, interrupted only by the downward-curved gash of her mouth and dark eyes. She was one of many, and during her lifetime she would be nicknamed Amelia by a scientist named Molly Lutcavage.

Many months earlier, back when she had hatched and grown into a three-millimeter-long larval fish, Amelia's eyes opened on the warm waters of the Mediterranean. Bluefin tuna spawn only when water reaches between 20 and 29 degrees Celsius. When they do, they breed in the dead of night, between 2 A.M. and 4 A.M., producing milky clouds of millions of eggs and sperm that drift five meters below the ocean's surface. Once fertilized, each egg measures about a millimeter across and will float on balmy currents for one to three days.

Amelia's first meal after hatching was an oil droplet contained in her own yolk sac. Her tiny body quickly developed huge black eyes and a digestive system, including a disproportionately large, hinged jaw with a toothsome underbite. Over the next two weeks she grew needle-sharp teeth, a stomach, and gastric glands to eat and digest crustaceans, including the copepods and water fleas that made up her earliest diet. When she struggled to find food, she ate smaller bluefin larvae to survive. By 25 days old, she had developed a swim bladder and a notochord, or early backbone, and started to swim with other centimeters-long bluefin her own size. They ate any fish smaller than themselves, evading predators and growing larger, eventually following their species' annual outward migration westward to the Atlantic's cold, nutrient-rich waters.

Sometime between her birth and 2004, Amelia accomplished her first feat of long-distance swimming, crossing the Atlantic Ocean's entire breadth before arriving off the coast of western Rhode Island. As she matured, she was on track to beat the odds, becoming one of only two of her fellow 30 million fertilized bluefin tuna eggs to make it to full-grown adulthood.

As Amelia cruised the waters off Rhode Island, it was a day like any other. But it was also a day for fishing.

A handful of kilometers away, on that same dark, predawn September 2004 morning, Al Anderson's hulking black truck sped along the paved road that dead-ended into Narragansett's Snug Harbor. Its headlights flickered between birch and alder, darkened homes and mailboxes, blurring them into stop-motion. Al had waved goodbye to his wife, Daryl, who flapped back from their porch with her dish towel, and he couldn't wait to get fishing. Driving down the steep pitch toward sea level, tall, anvil-headed Al, with a wiry mass of black hair and his trademark trimmed mustache, parked in a prime pull-through parking spot beside a drooping tree. He killed the gas and stepped down, eyes scanning the dock for competition, then walked toward his boat, casting lanky floodlight-lit shadows on the dock's gray, flaking planks. Dried-out iridescent scales glinted in the light under his feet, mirroring the fading stars above. There, in prime position near the diesel pump, rocked the *Prowler*, the third of Al's boats to bear the name.

A 13-meter sportfishing craft, this *Prowler* was bone white, her hull made by a North Carolina builder named Alex Willis. Back when Al had made the leap to fishing full-time, he had spent long days searching the Tar Heel State for the functional boat of his dreams—one he could afford, of course. North Carolina has long been renowned for its curled-lip ocean yachts, and Al knew the boat would likely be the biggest he'd ever own.

That morning in 2004, Al had been hired by charter clients to take them out fishing, which he did nearly every day between the spring and late fall when the weather cooperated. This day promised to be a beauty. Al's mate for the season, Bryan, was already on the boat, and their clients were set to show up at any minute.

Those clients—32-year-old Jason Williams and his brother, David, and father, Richard—woke up that morning on musty-smelling mattresses at a nearby motel. The place was "kind of a dump," Williams told me, but it was close to the harbor. He had contacted Al by telephone a few weeks earlier after reading an article penned by the longtime charter captain about catching bluefin on a lightweight fly rod. Back then, Williams spent every spare waking moment either going on or planning fly fishing trips, and in the fall of 2004, huge numbers of bluefin had shown up just west of Block Island, powering within a strip of water known as the Race. So Williams took the bait, convincing his father to cover the $400 deposit and pay Al's higher-than-average $1,350 group fee.

As they got out of their car and headed to the marina in darkness, Williams walked toward the ocean. The wind barely whispered, blowing at less than five knots from the southwest. From the dock, Williams caught a glimpse of Al standing in the *Prowler*'s wheelhouse. "Go ahead," Williams told his brother and father, who followed his instructions, striding down onto the pristine white of the *Prowler*'s deck.

Descending from his perch, Al moved toward them, his arms spread wide. Williams immediately knew they weren't being welcomed. Instead, they were being *pushed off the boat*. "He hadn't invited us on yet. He just came walking at us and shoved us back off," said Williams. "It wasn't like, 'Good morning.' It was like, 'All right, you guys get off the boat and I'll tell you when to get on the boat.'" Walking backward, the shocked fly fisherman thought, "Holy shit. This is going to be a long day." The moment hung in the air, the three clients standing awkwardly on the planks. A beat passed. Then another beat, and as if nothing had happened, Al smiled and genially waved the group aboard.

By 6:30 A.M. the *Prowler* had powered out of the harbor. Over the roar of its dual motors, Al yelled for Jason, who carefully picked his way toward the helm. To his relief, Al only wanted to chat fishing. Previous trips that week had gone well, the old salt boasted with enthusiasm, and he had high hopes for their trip. At one point, Williams caught himself staring at Al's face, lit red by the dim glowing lights of the boat's crimson instrumentation and cockpit lights. It felt more cozy than spooky and was an effect in a fishing cabin he had never seen before. The barely visible lighting made perfect sense; red light doesn't kill your night vision, and the low lighting signaled to Williams that this was a boat devoid, as he felt, of "bullshit." Obviously Al knew what he was doing, and had done it thousands of times before. And the *Prowler* was immaculate: there were no buckets lying around full of rusty fillet knives, no smears of desiccated fish blood left on the deck.

Less than an hour after leaving the harbor, they arrived at that day's fishing grounds, a popular spot called the Mud Hole. Al motored until he found a spot he liked, looking for fish on his sonar and carefully analyzing the currents that pushed around them. Once floating atop a patch he felt good about, Al killed the big engines, allowing the boat to drift as he and Bryan double-checked their rods and lines. That day, in addition to Williams fly fishing with his own gear, the group fished from standard rods and reels spooled with green monofilament, a single extruded nylon strand Al favored for its strength and near transparency underwater.

The lines' ends were tied with four-ounce diamond jigs, each with an elongated bullet shape and four light-catching sides with a hook on the end. "Diamonds" have been handcrafted by enterprising fishermen for more than a century; they are simply dropped into the water, where they fall vertically, flashing like a small fish when pulled behind a drifting boat or reeled in. They're often used to catch black sea bass and bluefish. Here he was, chuckled Williams to himself, trying to land a mighty marine predator with a rod a kid could lift and tackle he could buy at any basic tackle shop.

First, Al needed to let the tuna know where they were. To do that, he and Bryan chummed the water, dumping chunks of creamy-fleshed butterfish behind the boat. Within moments of their first line hitting the water, the Williams men started pulling at strikes on their rods, hauling juvenile bluefin tuna into the boat by the dozens, feeling the adrenaline hit as they hauled on the reels over and over again. They couldn't believe their luck. The one- and two-year-old fish hanging from their hooks were nicknamed "footballs" for their prolate spheroid shape and size but were denser than they looked and fought hard. Williams savored every second of the action. He felt the tips of the rods jerk in his hands with every strike, bracing his feet and stomach, sensing tension on the line as he fought each fish. The reel spun every time it ran from the boat, his rod tip bending like a weighted willow branch toward the water.

During his first phone conversation with Williams, Al had made himself clear: unlike most charter clients, their group would keep only one fish. Even if they could legally keep more, if it were up to Al, they would catch as many as they could and then tag and release them all. To be sure, most clients returned with wet bagfuls of fish to feed families and friends, to brag about at backyard barbecues over beers. But what Al really wanted to do out on the water was to mark as many fish as possible with tiny bits of numbered plastic, set those fish free, and wait for the data to flow in. And that particular day was on track to be one of his greatest yet for tuna tagging.

Two years earlier, Al had re-caught a bluefin tuna he had previously tagged himself, an experience that fanned the flames of his obsession. Daryl recorded the moment in Al's tidy annual logbook, a lined dossier with a black-and-white mottled cover and pages that crackled with ink running from margin to margin. With every year he fished, those logbooks piled up, their spines and contents bulging with successes and failures. Since his first days of tagging fish, Al once wrote in an unpublished memoir, he had long hoped to beat the extremely slim odds that he, a single fisherman, could catch the same fish twice.

"How naïve I was!" he wrote in 2018 of those early hopes, wryly noting that it "only" took 36 years of bluefin tagging for it to happen. On the occasion, officials at the National Marine Fisheries Service (NMFS) told him only a dedicated tagger could ever hope to recapture a bluefin they themselves had marked before. "The simple satisfaction of recycling this particular fish makes all the effort over the years worthwhile," he said. "You can't imagine how good that feels." Accomplishing a feat that few anglers could fulfilled a primal need in Al; he wanted to feel remarkable, and to know that through hard work, preparation, and grit he could accomplish something extraordinary.

That September day with the Williams family was the type of fishing day Al dreamed of: calm and sunny atop a gently rolling ocean that hadn't yet lost all its summer heat, with a seemingly unceasing well of fish eager to take the bait. After the first few fish, Jason Williams couldn't believe how many they were catching. "I'm going to sound like an idiot," he said as he and I discussed his hazy memories of that day, "but you'd get bored before you'd get tired." More than once, he and his brother or father were fighting a fish simultaneously on different rods as their tips arced toward the ocean's surface and each hard-fighting baby bluefin below.

To start the tagging process, first Al would scoop each tuna out of the water using a net designed for striped bass. (The comically small net struck Williams as ingenious. Imagine a mighty bluefin tuna in that itty-bitty net!) Moving smoothly to hold the slapping tuna steady, Al would use a swift strike from a tagging stick to insert a thin, round yellow spaghetti tag near the back of each fish's larger front dorsal fin.

After long moments spent gasping for air, each tagged fish needed to get back in the water as quickly as possible. With his strong right hand holding it firmly by the body, Al would drop it nose-first into the chilly water. Some fishermen simply toss a fish back overboard, but if it lands tail-first it can fall into shock and die. Those dead fish drift down past their schoolmates, the wasted charnel of human recreation. Later, standing at the *Prowler's* console, Al would officially record the weight and length of all the fish he caught on tagging data cards, which he'd then mail back to the NMFS.

"Few in the New England charter fishing industry have ever understood the advantages (immediate and longer range) to assuming their logical roles as stewards of the resource(s) on which their very livelihood depends," Al wrote in his 2012 book, *Island Stripers.* Over the years fellow skippers ridiculed him for his fish tagging, even taunting him for passing up a lucrative local black-market trade in selling undersized tuna to hungry Rhode Islanders. "If that was their attitude—if, that is, they measure a day's fishing against its yield in dead meat (much of it destined for freezer burn and, ultimately, the landfill)—I wanted no part of it (or them)."

NOT MANY PEOPLE in Al's life ever knew, but his brash outward persona had been shaped by a deeply painful childhood and a recurring fear of being abandoned. Memories from his early life haunted him and, according to those who knew him best, formed the foundation of his single-minded dedication to success that eventually sparked his fascination with fish tagging and bluefin tuna.

Al was born in 1938, sandwiched by the Great Depression and the Second World War, the only child of Arthur and Isabelle Anderson. Throughout Al's infancy and toddlerhood, his father couldn't find steady work and picked up side jobs for cash. His mother was prone to long, awkward silences and outbursts, and she struggled to hold down a job; sometimes she spoke in tongues or went on meaningless rants. Many in their New Jersey neighborhood whispered behind her back.

The year Al finished third grade, his parents abruptly separated. His father insisted on keeping Al but couldn't stay in his mother-in-law's home, where the family had lived for all of Al's early life. Suddenly homeless, Arthur packed up his son and a few small bags and the pair started sleeping in the family car, which he parked on the periphery of public parks and alongside abandoned beaches. After some weeks, Al caught a terrible cold that forced the pair to move into a big, drafty house owned by Al's paternal grandfather, a home already occupied by one of Arthur's brothers and two of Al's cousins. Father and son bunked together in an attic bedroom on the third floor, which, in the frigid, Atlantic-blown winters, struggled to catch heat from the basement's coal-burning furnace. In lieu of rent, Arthur's father tasked him with keeping the boiler lit; nine-year-old Al cleared the ashes every morning.

During that winter's coldest moments, Al dreamed of spring

and adventures of warmer weather. By six years old he had already fallen in love with fishing. He adored the restful rhythm of sitting by a pond, of dropping his line into the water and pulling up a fish. On Al's favorite, most hopeful days, his mother would pick him up from school in an old inherited Plymouth and they'd drive to nearby Fords Park. There they'd sit, side by side, eating sandwiches as Al fished with a stick, string, and a small ball of bread skewered on a bent pin. He caught nearly every sunfish in that pond.

With access to both saltwater ocean and freshwater lakes and ponds, most midcentury New Jerseyans could try to catch a fish for dinner when the urge struck. It was a time without recreational or commercial fishing limits, so the only true cap on one's catch was time, patience, equipment, and how many fish could be carried home. Before the arrival of factory freezers and deep-sea sonar, the ocean off North America's Atlantic coast teemed with life. A hundred years after John Cabot arrived on Indigenous-inhabited Turtle Island—the name given to North America by many of its First Nations—British fishermen reported schools of cod "so thick by the shore that we hardly have been able to row a boat through them." Just off the eastern seaboard, flocks of birds shadowed enormous runs of fish sprawling millions of cubic meters, chasing tiny, oily, ecosystem-sustaining fish like herring and mackerel as well as the giant tuna and sharks that pursued them.

In the years following the Second World War, steam- and diesel-powered ships swiftly replaced fishing under sail as the dominant technology, and shipbuilders equipped many of these new vessels with trawling and dragging nets that could catch thousands of fish at once. They scooped up fish from the water column across huge distances or scraped seafloor-dwelling fish, like flounder, off the ocean bottom. With some scientists startled by the high rate of extraction, this era also birthed the science of fisheries management, which was, essentially, the mathematical modeling and prediction of the health of aquatic species in proximate environments and ecosystems. In 1931, America's federally funded Woods Hole Oceanographic Institution (WHOI) commissioned its first vessel specially built for marine scientific research, the 43.5-meter-long sailing ship *Atlantis*. It was later joined by *Asterias*, a hard-pine commercial-style boat, and the *Albatross III*, a former fishing trawler. While aboard, WHOI researchers studied populations of groundfish like haddock, hake, and fluke offshore to better understand the valuable and heavily fished species.

In 1952, scientists with the US Fish and Wildlife Service's technological laboratory in Boston used an experimental trawler named *Delaware* to invent a technology that could freeze fish at sea. The technique offered a "practical solution to the dream of deep-sea fishermen," declared American fisheries technologist Harris W. Magnusson at a 1953 conference in Stockholm, "to deliver a capacity load with . . . uniformly high quality and . . . the highest prices." This new invention relied on a rotating system of 11 round wire-mesh baskets, each 2 meters long and 0.6 meters in diameter, that rotated through an ice-cold tank of salted brine. Each basket held up to 500 pounds of freshly caught fish, and the system could freeze up to 770 pounds of fish solid every hour, which were then slid, via a chute, into the trawler's 100,000 liters of cold storage—about the same volume as 10 cement trucks. Once workers transferred the fish off the vessel into canvas baskets used at Boston's bustling fish pier, they were held in a warehouse and eventually thawed in vats of fresh water. Filleted in factories, the previously frozen fish produced "uniform, high-quality, marketable, and delectable filet[s]," wrote Magnusson. Thanks to this new process, entrepreneurs could catch and store fish and ship them to towns and cities hours away. The change represented an upheaval for fishermen who, since Cabot's time, had been forced to rely on either selling fresh fish locally or preserving fish using time- and labor-intensive techniques like canning, smoking, salting, or drying.

As Al's family struggled to keep food on the table, the young angler quickly learned that he could help with dinner and earn praise on the side by catching fish. After more than a year spent living with his father in the drafty attic, 11-year-old Al moved back in with his mother, who had found secretarial work, and moved with her to nearby Metuchen. Every day that first summer, he hopped on his bike and pedaled 2.5 kilometers by himself to Roosevelt Park to fish in its lake. Going out in the morning, at noon, and even sometimes after dark, Al observed the rhythms of each species, the food they liked, the behaviors that drew them to his bait. He caught perch, the occasional largemouth bass, and at least a few weeks' worth of early-season stocked trout every spring. When he finished fishing for the day, he hid his crude, cheap tackle in the lakeside scrub and pedaled home, bearing the best, plumpest fish for his mother to cook for dinner.

In the 1950s, Fords, New Jersey, sat at the crux of two newly built superhighways—the New Jersey Turnpike and the Garden State Parkway—and had recently started to reap the rewards of the accompanying boom. By the middle of the decade, Al's beloved uncle Harry Bernau had landed a plum position collecting tolls on the parkway. In their spare time, Harry and his wife, Lillian, ran a small home business making and repairing saltwater fishing rods in their basement. From the sidelines of the Bernaus' workshop, Al would watch, rapt, as his uncle bound fiberglass poles with thread, firmly tying metal line guides onto the rods with tight loops. Lillian, with her steady hand, then coated the thread with layers of clear lacquer to hold the line guides fast. In that basement, Harry built his own metal lures, designed to dip and wobble in the surf to attract the attention of curious striped bass. When Al finally convinced his uncle of his serious interest in fishing, Harry helped him buy a new state-of-the-art spinning reel, a black, riveted device with an embossed Garcia Mitchell logo in curling script.

As Al found solace with his uncle during those troubled years, his extended family began to pressure his father, Arthur, to make a fresh start for the sake of his mentally ill wife and his sidelined son: it was time for him to take responsibility, they lectured. Eventually, Arthur caved and made a plan. One afternoon, he picked up Isabelle and Al, then 14, and drove them back to Safran Avenue, the street where Al had been born, to proudly tour them around a vacant lot. Al found it unappealing and scrubby, a postage stamp of land filled with large trees along a trickling creek. But Arthur had a surprise for Al: he planned to build a house there for the three of them, with the help of his brothers. This was also how Al learned his parents were reuniting.

With their newly re-formed family unit on a tight budget, Al often worked as his father's only helper. He watched as a backhoe dug the foundation. His father set up string lines for its poured concrete slab and used cinder blocks for its basement walls. Al worked alongside his father and uncles, learning how to use a saw and finish a joint—and that nothing got done unless you did it yourself. Al took that last lesson deeply to heart.

One blustery fall afternoon, Uncle Harry's brand-new black Mercury sedan rolled to a stop in front of Al's home. Fords was still in its postwar boom, fueled by demand for labor and materials from nearby New York City, and new houses like Al's were popping up like weeds in sidewalk cracks. Reaching his destination, Harry stepped out onto the sodden grass in front of a small bungalow. He scanned the house as Al, by then a tall, lanky teen with wavy black hair, opened the front door and made his way down its front steps.

After they'd greeted each other, Harry stood back and opened the Mercury's smiling black trunk. With a clunk and a swing, Harry revealed his prize: three fat striped bass, so fresh they were still moist. The fish, all as long as a man's arm, wobbled on opened sheets of newsprint, their bulbous eyes luminous and clear. Al's eyes widened. Harry had caught them that morning in the Sandy Hook surf, about an hour's drive south, using a thread-wrapped rod and curved metal lure, gear he had hand-built and hammered himself in his basement workshop. The teen drank in his uncle's pride. The showy car, the gleaming fish, the braggadocio—all of it dazzled. "Boy, oh boy," he thought. The glamour and accomplishment cast a spell over Al, hinting at a thrill he'd chase all his life.

~~~

In 1958, at the age of 20, Al enrolled in New Jersey's Fairleigh Dickinson University, breathing a sigh of relief at the chance to leave his parents' emotionally chilly home. When he left, he also packed his rod and tackle box. One day during his freshman year, Al spent a morning fishing trout with a friend in the nearby Hackensack River and lost himself in the reverie of casting and reeling. When he realized he was late for class, he sped back to school in his secondhand Chevy. Bursting into his biology class-room, he sheepishly pocketed himself into his seat with a squelch—he was still wearing chest-high rubber waders and a fishing vest. After class he gutted a fresh sackful of trout in the classroom's laboratory sink.

Al approached making friends strategically. In his sophomore year at Fairleigh Dickinson he pledged to the Epsilon Nu Delta fraternity, a less-popular group of lower-income aspirants who had been passed over by other Greek clubs. (Some called the group "the Poor Fellows.") During pledging, one fraternity brother drove and dropped Al and two other students in the boonies of New Jersey, only for Al to realize they were near a familiar trout hatchery. He led the small group to a nearby diner, where they thumbed a ride back to campus in the cab of a tractor-trailer and returned in record time. Al's peers elected him senior class and fra-ternity president, and his numerous athletics trophies—basketball, fast-pitch, cross-country—jammed the top of his dormitory dresser. In his senior year, Al met and courted his first serious girlfriend, a woman named Bonnie, while he worked as a clothes model and stock car driver for hire at the local quarter-mile racetrack.

For a year after graduating in 1960, Al worked in a laboratory at the University at Buffalo's school of medicine. He spent hours cleaning glassware, eventually working his way up to running the lab's then-state-of-the-art ultracentrifuge. Every weekend, he drove his Chevy the 600-mile round-trip back to New Jersey to visit Bonnie. He continued to fish, one time winning a university contest for trout fishing in the Delaware River. In 1961, he was offered a graduate assistantship in fish parasitology at Long Island's Adelphi University, and he and Bonnie married that summer. By his own account, he left the University at Buffalo a "cum laude graduate of wild brook trout fishing."

Within months of tying the knot, Al and his wife received a letter from the Nassau County sheriff's office regarding a string of tampered mailboxes at apartment complexes. In the letter's aftermath, Al realized Bonnie had spent all the money he had given her monthly for rent and that their landlord had never been paid, nor had several of their credit card accounts. Bonnie skipped town in his car, leaving Al deep in debt, heartbroken, and living in an apartment he could barely afford. For weeks, carless and broke, he hitchhiked to campus. It was at this juncture of his life Al started the unusual fishing practice that became his signature.

Starting in 1961, a newly single Al often risked missing his morning classes to catch a few largemouth bass in Long Island's Lake Ronkonkoma. Standing on its sandy beaches, the lanky, square-jawed young man squinted into the sun past his battered baseball cap as he cast his line. He always tried to arrive at the lake when it was still dark, when the incoming day barely tickled the horizon in purple and mauve. It was the best time, he knew, because it was the best time to catch fish.

He picked a prime stretch of beach, setting down his carefully organized tackle box and the prized Mitchell Garcia rod and reel his uncle had given him a decade before. His leather loafers shifted on the fine sand as he pitched his line toward a sunken trailer that lay on the lake's dark bottom. The trailer's corroded nooks and crannies provided perfect cover for the schooling largemouth bass that darted among the gooey algae and floating candy wrappers. The bass chased minnows and leapt at insects that hovered too close to the surface, and were, as he joked, "highly cooperative" and easy to catch. He tempted the school's ancient feeding instincts with his lures, getting their attention with short, jerking movements from his wrist, his rod's light linen line flashing in the growing morning light as he pulled in fish after fish.

One day, he brought a spool of copper wire to the lake to attempt something new.

Once a bass took his hook and he had drawn it close to shore using his rod, Al lifted it out of the water with a fine-meshed net; the finer weave was gentler on the bass's delicate skin and made it easier to scoop up its slippery, forest-green body. On his hands he wore thin, tight-fitting white cotton gloves to improve his grip and preserve the fish's veil of protective slime. Clenching the shimmering, flopping bass firmly with his left hand, its flesh bulging slightly under his long fingers, he used his right hand to pick up a straight, thin, finger-length scrap of copper wire he had precut off the spool. With a downward poke he jammed the wire's sharpest end through the upper back of each fish, near its dorsal fin, as the fish kicked against the sensation. He quickly twisted the open ends of the wire closed and released the bass back into the water. It was the first fish he ever tagged.

For months, Al fished near the sunken trailer, catching and tagging dozens of largemouth bass, using the same technique. One day he realized that he had tagged the same largemouth bass six times—as he pulled the flapping fish from the water, sunlight flashed on five previously inserted copper wire tags poking from its back. Despite its crown of metal, Al marveled, the bass seemed to be perfectly healthy and unaffected by the multiple perforations; it still fought his rod vigorously each time and the multiple tagging holes had healed tightly around the copper wire. Touching their sharp tips before releasing the bass yet again, Al found himself entranced by what those multiple tags indicated: this was a small but healthy population of bass, in balance with the man who was fishing from it. He didn't worry much about whether his tagging hurt the fish: at the time, popular understanding of animal cruelty extended only to specific mammals such as cows, bears, and dogs, and to chickens. Fish, without facial expressions or the ability to make noise above water, remained in a lesser category.[1]

A personal interest in tagging galvanized Al's fishing, but by 1964 he still hadn't figured out how to dovetail his love of fishing with earning a living. Wanting to be close to the ocean, he applied and was accepted into a PhD program at the University of Rhode Island. There he befriended one of the school's prominent fish scientists, ichthyologist William H. Krueger, and bought a leaky 14-foot-long Bristol skiff with a 20-horsepower Evinrude outboard motor he towed on a small trailer. During the day Al researched *Austrobilharzia variglandis*, the parasite that causes clam-digger's itch, and studied his dense copy of *Chapman Navigation Rules* late into the night. Regardless of where his academic career went, he decided, he would eventually become a licensed charter captain.

In 1967, Al started tagging striped bass using tags purchased from the American Littoral Society (ALS), a New Jersey–based environmental group dedicated to promoting the study and conservation of marine life and habitat in the US Northeast. ("Littoral" means "near the coast.") Founded two years prior, the ALS tagging program was one of the first recreational tagging programs in the United States, although research tags had been deployed by scientists there since 1873. Al fumbled and splashed around in those earliest formal tagging attempts, struggles that reminded him of the slapdash antics in the Marx Brothers' film *Mixed Nuts*. (It was this nickname he gave his and others' first tagging expeditions and a name he later gave his first five-meter Boston Whaler.) But eventually he figured out a smooth system for tagging stripers and grew to love the work, announcing to incredulous charter clients and anyone else on the docks within earshot that the society's goal was to turn fishermen into scientists. Still, he took care to ensure clients' coolers brimmed with fish before suggesting catch-and-release. Most weren't interested at first—what was the point of paying hundreds of dollars for a single fishing trip if you couldn't even eat what you caught?—but Al worked on selling the idea to fellow fish enthusiasts.

He was also attracting a new wave of clients interested in catching Atlantic bluefin tuna, a warm-bodied fish with a reputation as a hard fighter that schooled off Rhode Island's coast by the thousands.

In the 1960s, recreational anglers increasingly headed to Block Island, Al Anderson's favorite fishing hole, in search of big bluefin. Most giant bluefin in the region averaged between 280 and 600 pounds, although bluefin off the Canadian province of Nova Scotia, some gossiped, grew many times larger. In his earliest angling days off Rhode Island, Al had cursed the numerous schools of small bluefin tuna that blasted past his striped bass lures. Sometimes they stayed far off the coast; on other, rarer days, bystanders on New England's coastal beaches could see them leaping offshore. They were a pain in the neck, those fast, powerful fish that took his lures and broke his lines. But they were also fun to catch.

The fish's speed intrigued marine scientists like Frank Carey, a Woods Hole Oceanographic Institution biologist who, working alongside fellow researchers, first identified the mechanisms behind the bluefin's warm-bloodedness in 1966. The species' ability to regulate its internal body temperature manifested in powerful ways, including the bluefin's ability to preserve the warm tone of its muscles, and therefore speed and strength, even when swimming in chilly waters. French oceanographer Jacques Cousteau once described the experience of swimming alongside a school of 60 giant Atlantic bluefin tuna in the Mediterranean as fishermen hauled in the net in which they were trapped. As the net shrank to a third of its former size, the tuna exploded into a frenzy. "With the seeming momentum of locomotives, the tuna drove at me, head-on, obliquely, crosswise. It was out of the question for me to dodge them," he wrote in *The Silent World*. "Frightened out of sense of time, I . . . surfaced amidst the thrashing bodies. There was not a mark on my body. Even while running amok the giant fish had avoided me by inches, merely massaging me with back-wash when they sped past."

While the bluefin's precise vision and powerful speed were the culmination of millions of years of evolution, they were no match for the increasingly industrial capabilities of global fishing fleets. For decades starting in the postwar years, longliners and similarly high-capacity purse seiners had arrived on American and Canadian coastlines from all over the world, from Japan, Germany, Spain, and other countries, to catch bluefin en masse.[2] When they set lines or corralled their purse-shaped nets tight around giant schools of fish, their crews and captains often weren't picky about the size of fish they caught. These ship's crews primarily pursued smaller bluefin, as customers preferred the fish's medium-pink flesh. Sometimes they even threw away any truly giant fish or "granders"—fish weighing more than 1,000 pounds—because their blood-flushed red-purple flesh couldn't fetch the same price per pound at the canneries dotting the Atlantic coast. They preferred the lighter-colored meat of the albacore tuna, which was widely used in the tuna casseroles and vegetable-specked nouveau salads that adorned mid-century party spreads. For years, the market delegated darker bluefin as a poverty fish, good only for cats and Italian immigrants who could be convinced to buy it canned and cheap by the box.

Purse seiners often deeply bruised the fish they caught, crushing their small bodies under the weight of so many tonnes of creatures—nearly any ocean animal you could imagine—pulled together and dragged aboard the vessels. In these writhing, multi-tonne nets, countless bluefin died by suffocation. Their raw, venal gills pulsed open and closed as their blood oxygen levels dropped and eventually their hearts stopped. At times, there were instances where a ship needed only, say, 7 tonnes of tuna. If its crew caught 90 tonnes, that meant 83 tonnes tossed overboard. In the early 1960s, "several generations of tuna were wiped right out," Al once wrote, while accidental breakage of nets often killed hundreds of tuna at a time. It was a meaningless, pointless death (or the efficient disposal of bycatch, as far as many captains and their boats' shareholders were concerned), but in either case the waste was considered a necessary side effect. Sacrifices must be made, the reasoning went, to fill the maw of the world's newly emerged global fishing fleet and its customers.

Back then, the value of bluefin flesh hovered around five cents per pound, far less than bass or trout. That meant fishermen had limited options when it came to getting rid of their tuna catches: "You would either take a picture and let the fish go," Charlie Donilon, a Rhode Island captain and tuna industry veteran, told me, "or you'd bring it back to the dock, hang it up, take a picture with it dead, drop it back in your boat, and then go outside the break wall and dump it." A third option involved paying a truck driver to haul the giant fish to New York City. But if the fish sold for less than the driver's time plus fuel, the fisherman owed a debt for a fish no one wanted.

Still, during bluefin tournament time, it wasn't unusual to see a gape-mouthed giant tuna tied to a car's roof, displayed like a trophy buck during hunting season. Other times, fishermen would try to transport bluefin home in their trunks, which made their cars sag so deeply that their exhaust systems were in danger of bumping off on the rural Narragansett roads. "The odor of fish pervaded the car's interior for months to come," Al once observed, "forcing the sale of a vehicle in one known instance." There was something about those mighty fish that captured imaginations. For Al, it was the start of a fixation that would last until his final days. But the biggest bluefin rush had yet to come, and he had already positioned himself at the heart of a fishery for what was set to become the world's most valuable fish.

CHAPTER TWO

"YOU DAMNED WELL BETTER TAG"

~~~~~~~~~~~~~~~~~~~~~~~~~~~~~~~~~~

Al, 1960s

Who has known the ocean? Neither you nor I,

with our earth-bound senses.

—RACHEL CARSON, "UNDERSEA"

Putting out in his first *Prowler* on New England's waters throughout the 1960s, Al Anderson marveled at the Atlantic bluefin's theatrics as they leapt in the greasy slick of chum alongside his boat's gunwales. As he got better at finding and catching bluefin, he found himself increasingly obsessed with gathering data and learning about the powerful fish, including how they grew, moved, and migrated, and where they went after disappearing under the gunmetal blue chop.

Physiologically, a bluefin's body is nearly pure muscle. Its fins can be retracted into shaped hollows along its body, and its eyes sit flush to its head. Scientists have speculated that the yellow spiked finlets along its back and belly help reduce turbulence in the water at high speeds. First diagramed in 1978, the bluefin's "thunniform" swimming technique involves little, if any, head movement. That, paired with strong, rapid back-and-forth kicks from a powerful jointed tail, allow the fish to maintain high speeds for long periods of time. Growing up underwater, a bluefin like Amelia would have schooled with bluefin approximately her same size, not with a family group, as she hunted and rested. The position she held in the school would have shifted often and changed with the currents and circumstance: tuna often swim in soldier formation, moving side by side in a single horizontal row at the same depth, but have also been observed from the air in spinning-wheel-like formations and densely packed domes at the ocean's surface. "Sometimes schools of bluefin swim under the curl of a wave, stacked like a flock of birds," Douglas Whynott writes in his book *Giant Bluefin*, "as though trying to get a look at the upper world from an aquarium window."

For a fish with very little color, a bluefin's typically muted skin has an otherworldly capacity to shift into a rainbow of colors—silver, gray, black but also blue, purple, orange—that shift and shimmer. Like octopus and squid, the species has skin that evolved with pigment-producing chromatophore cells controlled by the animal's central nervous system. In a way, a bluefin's colors are like a tree's leaves in autumn: as a tuna fights and those cells expand and contract, its skin puts on a vibrant display fishermen call "flashing." While catching and tagging, Al sometimes found himself flabbergasted by the fish's beauty as its skin morphed before his eyes, turning spotted, or variegated, or even striped dark and light like a zebra's coat.

Omnivorous in his search for better ways to tag fish as he had first done in that fall of 1961 on Long Island, Al sat at his small desk and immersed himself in an emerging wave of scientific research into bluefin tuna, marking up magazine articles and scientific studies with scrawls in the margins and pinning them up in his office. Many of those academic papers had been written by a Massachusetts-based scientist named Frank J. Mather III. Born in 1911, Mather was a hobby fisherman and well-to-do son of a famous art historian. He built his career on a class of ocean-spanning fish called pelagics—a term drawn from oceanographers' name for the mid-ocean region between the sea's floor and surface, which is, by volume, the largest habitat on Earth. Bluefin tuna are pelagic, as are marlin and swordfish, and for Mather they were the ultimate catch. During the Second World War, Mather worked in New York City as a naval architect for the US Navy. After the war, he pursued his teenage dream of studying fish and lucked into an early career position as a research associate at Woods Hole Oceanographic Institution simply by being one of the few showing interest in the field.

At the time, marine scientists knew very little about bluefin— where they spawned and migrated, or how they grew or pursued prey. There were large concentrations of the fish off Nova Scotia and Bermuda, the North Sea and the Mediterranean, but their lives between those points were largely a mystery. As of the mid-1950s, most international government-funded fisheries research, according to Mather, focused on how to exploit big fish as an industrial resource. Consequently, as Mather wrote, their programs were aimed "chiefly toward discovering new tuna fishing grounds, and toward determining how to regulate fishing so as to get out of tuna resources the highest sustainable harvest." Filling this gap between commercial, game, and research fishing, Mather proposed, would be his Cooperative Game Fish Tagging Program, a first-of-its-kind research program that would work with recreational fishermen along the Atlantic coast to catch and track tuna and other saltwater giants.

"No one has yet succeeded in tracing the migrations of any of the great game fishes by tagging," he wrote. "If anglers believe it important to learn about these species on which their recreation depends—and they should, for the purposes of conservation . . . — the necessary research will have to be done almost entirely by privately endowed marine laboratories." In requesting a $50,000 budget (half a million dollars today), Mather transformed his weekend hobby into serious scientific research. At first, many northeastern fishermen rejected the idea of tagging bluefin, some thinking that Mather's studies would only help other fishermen, particularly those in other regions of the Atlantic, catch more of "their" fish. But others were more easily persuaded. "The only way we are going to learn anything about the fish is through tagging," Mather once told Bob Linton, a well-known boatbuilder and charter fisherman who fished offshore for tuna throughout the late 1950s and early 1960s. Thanks to Mather, Linton's curiosity outweighed his trepidation and he became one of the first charter fishermen to tag for the researcher. The scientist's undeniable logic, as Linton once told Al, "made a lot of sense."

Mather, along with a biologist named Howard Schuck, tagged his first tuna by inserting a stamped, numbered hook through the fish's jaw off the coast of Bimini in the Bahamas in May 1952. (The early technique was based on metal ear tags attached to cattle.) That fall the two scientists went "through the roof" with astonishment when a Nova Scotia tuna trap owner called Schuck with news of the fish's recapture thousands of kilometers north from where it had been tagged. Within a few years, data was rolling in from around the world as fishermen caught and reported the recapture of Mather-tagged bluefin tuna. In 1962, one fish tagged off the Bahamas migrated nearly 10,000 kilometers in 50 days and was caught off Norway's coast. That fish moved at such high speeds that it traveled a distance of nearly five consecutive marathons every day for more than a month.

The simple science of counting and recording fish tagging data has been practiced sporadically in Europe since the mid-1800s, when curious Scottish landowners tagged salmon and trout running through their properties to see where they ended up. Before that, Indigenous and early European fishers engaged in a more informal version of the practice, recognizing and drawing meaning from the rediscovery of their own and others' handmade hooks lodged in the mouths of re-caught fish.

By marking a fish and releasing it, modern scientists can plant an irrefutable data point, a record of that one fish caught in a particular place at an exact size and weight. If a fish tagger catches and tags 100 more fish, the likelihood that a tagged fish will be re-caught goes up. When that happens, it then becomes possible to plot where a single fish has traveled and grown over time. Tag 1,000 more fish, and the confluence of data creates a map bristling with meaning, packed with information on where fish are born, where they die, how fast and where they travel, and whether their communities are healthy or at risk of being wiped out.

Imagine you have 100 apples in a pond, and only 1 has a sticker on it: the likelihood of dropping a net into the water and scooping up the one with the sticker is low. But if there are 5 apples left in the pond and only 1 has a sticker, and it's netted repeatedly, then that's statistical evidence of fewer apples in the pond. If many thousands of bass lived in Lake Ronkonkoma—Long Island's deepest and largest lake—it was unlikely Al would catch the same fish repeatedly. Yet there he had stood, casting toward that sunken trailer, catching and freeing the same fish again and again. The recatching itself had meaning: it meant he was fishing into a small, relatively static population, so if he had caught and kept every fish, the population could disappear. Instead, thrilled by the discovery, he continued to set most of them free.

In 1967, after years of hearing about Mather's work on the docks, Al and Mather connected and the two started corresponding by post, trading pleasantries and swapping data and ideas. Mather asked if Al would contribute to his tagging efforts, and the fisherman felt honored to be chosen as a collaborator. Some of the earliest devices Mather, Al, and other early taggers used to mark tuna were modeled on shark tags, which had stainless steel tabs that expanded inside the fish to hold the tag in place. They easily stayed lodged in slower-moving sharks, but in bluefin, which reach speeds of up to 65 kilometers an hour, friction caused by high speed could pop the metal tabs right off. As they collaborated on new, better ways to tag bluefin, Al helped Mather develop and prototype a plastic dart, with a similar silhouette to the expanding plastic anchors used to hold screws in a wall, that embedded and locked a few inches deep into a bluefin's flesh.

Once a tuna is caught, the most critical aspect of tagging is keeping the fish alive. Water must pass constantly over a bluefin's gills or it will suffocate, so each fish has to be tagged extremely quickly. Most taggers use a tuna tagging stick—essentially a wooden broom handle with a steel spike on the end—to insert a tag in a tuna's back, near its front dorsal fin. When dragging and tagging a fish off the side of a boat, which many captains prefer, a long stick works best; for captains who prefer to land and tag fish onboard their boats, a shorter sawed-off stick works better. Some even pump salt water into a fish's mouth and past its gills to make sure it's getting enough oxygen. In the latter case, Mather encouraged fishermen to use wet gloves when taking a fish out of the water, to avoid picking it up by the tail, and to work as quickly as possible to get the fish back in the ocean once it had been tagged.

The next most important element is the collection and recording of data, including the date, the GPS coordinates, how long the fight with the fish lasted, and the fish's length, as well as names and contact information for either the angler, the captain, or both. In filling out these tiny cards, Al's ability to notice any small detail awry echoed satisfyingly. As his father had taught him on his family's childhood construction site, details were important, and Al regarded those hundreds of forms as part of his personal scientific journey: every fact mattered.

In his magazine articles, Al extolled the satisfaction and karmic virtues of bluefin tagging, which, he always said, started with a fisherman gathering supplies: tagging cards, tags, a tagging stick, and gloves for grip and to prevent damaging the fish's skin. Once an angler hooked a fish, Al recommended letting it "catch its breath" in the water alongside the boat, especially after a hard fight. Using a wet glove or a snare, he'd hoist each fish aboard by the head and tail, laying it beside his carefully arrayed materials and tape measure. Once he inserted the tag and recorded the fish's length and weight, he'd slip the fish into the water headfirst, letting oxygen flow into its mouth and across its gills before setting it free. "Mather was a strange duck," Al told a magazine reporter, reflecting on their tagging work decades later. "[But] when you made a commitment to tag for him you damned well better tag."

LONG BEFORE YOUNG Al found refuge from his problems in recreational angling, long before even the dawn of industrialized fishing, the earliest human coastal communities relied on the sea to survive. The practices, failures, and smokehouse-filling successes of our ancestors were the earliest form of hypothesis testing on the ocean's rhythms and bounties: for those living inland, sourcing ocean-caught fish for a meal was difficult and expensive, and any producer transporting fisheries' products inland was forced to preserve them—either by salting, drying, smoking, fermenting, or using a combination of those techniques—before risking a long journey and a wasted shipment. Early coastal peoples depended on what they harvested from the sea, but the quick perishability of those products and localized catches meant that catching and eating fish remained intensely regional.

To succeed, early humans had to adapt to their environments. Where fish were plentiful or ran past communities on tides, stone weirs and woven nets often were the most successful. Where fish were large and could be cornered or baited with lures mimicking insects, harpoons or baited hooks worked better. Archeologists have found fish spears more than 90,000 years old and fishhooks 20,000 years old. Some of the earliest human communities, like those living in modern-day East Timor in Southeast Asia, depended heavily on fishing for food during an age when ice still encased most of North America. Stone sinkers weighted down land-based nets in modern-day Russia between 10,000 and 8,000 years ago—a period when cheese and winemaking first developed—while carvings more than 4,000 years old depict ancient Egyptians fishing with lines and nets from papyrus boats.

The ancient Phoenicians, whose culture peaked between 1100 and 200 BCE, originated many fish-catching technologies and techniques that were later widely used across northern Africa, southern Europe, and eventually North America and the world's burgeoning colonial outposts. By the time Al started fishing, these technologies had evolved to use modern materials and manufacturing, but all would have been recognizable to the ancients. There are, after all, only so many ways to pull fish out of water efficiently. Notably, this included the Phoenicians' use of weighted setnets and seines—deep nets gathered into a semicircle against a beach—to corral tuna as they migrated. Previously, Indigenous tribes in North America's Pacific Northwest coast used similar innovations, weaving nets from nettle fibers, spruce roots, willow, and cedar bark to capture the region's pulsing runs of salmon. For both these groups, larger catches made possible from the labor of fewer people meant an escape from hunger and later poverty, especially when paired with early preservation techniques like salting and smoking. Rich in fatty acids and protein, fish provided sustenance and certainty in largely uncertain times and came to form the nutritional backbone of many ancient societies.

Waves of technological advancements transformed a way of life into an industry. For example, the precursor to today's modern propeller-driven trawler, a ship that catches fish by trawling the ocean floor that can devastate swaths of healthy ecosystems, was a Dutch vessel called a *dogger*. Builders modeled the square-sailed late-1600s vessel, which could carry about a tonne of bait and three tonnes of salt, on the sturdy remains of Norse ships. Doggers also carried a half tonne of food and a half tonne of firewood to keep the crew warm and fed on the frigid North Sea.

This scaling up of technology and mechanization troubled British fishermen as early as 1376, when one group submitted a petition to their country's Parliament objecting to the use of a new device dubbed a *wondyrechaun*. Their formerly "plenteous fishing," they contended, had been wiped out by fishermen using destructive gear towed along the seafloor: a wooden beam three meters wide towing a five-meter-long net weighted with lead-threaded ropes and stones. That government eventually introduced bans and restrictions on the devastating devices, and two British fishermen were executed in 1583 for using metal chains on their trawls. Still, ships under sail pulled trawls throughout the 1800s, and in 1863 one witness told a British royal commission charged with investigating the practice that trawling, in fact, helped the ecosystem and provided more food for fish. It did, in a way, as dead and dying creatures left in the wake of a first trawler attracted waves of opportunistic scavengers that were often caught in a subsequent trawl.

It was especially challenging for early scientists to learn about the ocean, which was harder to study than a forest or a desert, given the tools and technologies of the time. Until the mid-20th century, they largely relied on catch statistics gathered from fishermen and were forced to piece together the rest with observation and argument. The ocean held us at arm's length for centuries, and in the absence of good information to prove otherwise, many humans simply treated the ecosystem and its creatures as if both were infinite.

These patterns also extended to colonial expansion. As settlers arrived in North America, they swiftly ran roughshod over sustainable Indigenous fisheries management practices that had kept salmon, crab, and groundfish populations like cod, haddock, and pollock healthy for generations. Within two decades of British rule in the Canadian province of British Columbia, for example, the epic, splashing runs of huge coho and Chinook salmon up Pacific Northwest rivers had been decimated. Some of the earliest fisheries data collected by colonial governments were landing records of how many fish had been caught annually and where. They didn't do this to protect species; they were politically motivated attempts to control and profit off the growing industry. The ocean was the limit, but the ocean was limitless.

GROUPS OF FISH have historically been called "stocks," parlance for a manageable resource of domesticated animals such as cattle or other farm animals, and the conservation of fish stocks first emerged as a political issue as early as the 16th century. Since medieval times, communities, regions, and guilds enacted laws or made agreements with other bodies to regulate gear, fishing location, and methods. Formal bodies instituted to regulate fishing have existed in countries including Japan, France, and Spain since the 17th century. But with a shift toward capitalist and colonial extraction methods, and as the available technology to catch fish and travel the seas improved, by the 1800s it had become clear that the number of fish pulled from the ocean would continue to increase exponentially.

These huge, heaving fish populations, once Europeans became aware of them, were fated to be pursued in bigger and increasingly intense ways. In the late 1800s, a Scottish shipbuilder invented the steam-powered trawler, which, given its faster speed and potentially heavier load capacity, could outstrip the range and carrying capacity of the best sail-powered equivalent. Steam's heyday remained brief, however, and in 1910, French boat designer Benjamin Bénéteau launched the first motorized sardine fishing boat, the *Vainqueur des Jaloux*, which, roughly translated, means "conqueror of my haters." The vessel briefly caused outrage in nearby fishing communities that accused the "oil boat" of scaring fish away, but, as with most new technologies, fishermen quickly adopted the efficient design on both sides of the Atlantic.

Between 1901 and 1945, worldwide catches of fish every year reached 15 million metric tonnes, or the weight of more than 140,000 blue whales. Responding to this booming supply, a new wave of entrepreneurs and technologists arrived in the fishing industry throughout the early 1900s. On Terminal Island, home of Los Angeles's new railway terminal, the California Fish Company produced America's first canned tuna in 1903. It marketed the meat as a passable replacement for chicken, and the island's community of Japanese immigrants proved themselves to be adept at catching albacore and bluefin tuna. By 1907, there were nearly 600 Japanese fishermen working from the island. "The supply of tuna is inexhaustible," declared the son of the company's founder.

It was at this moment in human history that the very idea of catching fish and selling them to consumers started conforming to suit the needs of the globalizing industry. Fish in volume can feel inconceivable—speaking from personal experience, even a small commercial net full of thrashing fish sounds like the deafening hammer of torrential rain on a tin roof. As ambitious entrepreneurs would discover, fish sold in volume were easily commodified, and could be reframed by business vernacular into flattened economic units. And so capitalism again transmuted once-living creatures into a product ready for human consumption.

Innovation at sea only accelerated the change. In 1954, an 85-meter-long ship called *Fairtry* sailed the Atlantic for the first time. Designed with a hauling mechanism at its stern and an onboard fish-processing factory, the trawler could lift 55 tonnes of fish in a single catch.[1] Across all these technologies, the fact that commercial fisheries had become a race to the literal bottom seemed undeniable, even at a time when humans considered the ocean largely inexhaustible. More fishermen fishing from more boats were pulling more fish from the ocean, so boats needed to travel farther afield to catch the same numbers of fish. That meant more attention paid to species that schooled in nearby waters and could be sold at prices worth a day's fuel: species like bluefin tuna. By 1954, the United States had become both the world's largest producer and consumer of canned tuna, with every American eating nearly two pounds of canned tuna on average that year.

Between 1961 and 1965, Japanese longliners tripled their annual catch of bluefin, from 10,000 to 30,000 giant fish, to respond to domestic consumer demand. As Japanese diners' appetite for bluefin grew unchecked, and the price it commanded at Tokyo's Tsukiji market spiked, fishermen fished the closely related *Thunnus orientalis* with increasing fervor and it began disappearing from Japanese waters. That meant Japan's factory ships had to travel farther afield to fill the market's slick early-morning auction floors. Some of those ships ended up on the shores of the western Atlantic, where fishing for bluefin remained, at that time, a free-for-all.

The historic relationship between North American sportfishermen and giant bluefin was one that Japanese fishermen had heard about, but until populations of their own Pacific bluefin started to drop, making the long voyage to North America wasn't worth the effort. But by the late 1960s the economics of the fishery meant branching out. For middlemen who wanted to buy fish from American and Canadian fishermen, it helped that bluefin were being sold off the boat at laughably low prices. It was an appetite that established and fueled the global slaughter of bluefin that, by the following decade, changed the fishery forever.

OUT ON THE water throughout the 1960s and 1970s, Al found himself dodging the looping paths of giant purse seiners and trawlers. What they were engaged in could technically be called fishing, but to Al, what he did and what they did were so far apart as to be nearly unrelated. He was a mariner of experience and instinct, analyzing and gathering data, getting to know by heart the individual rips and currents of the waters off Block Island. In his daily fishing log, he kept detailed notes on wind speed and direction, temperature and time, recording both his daily plan and the rhythm of catches, and where and how deep fish had been found. With a disquiet he couldn't quite place, he watched huge fishing vessels cruise along the horizon, forced to only imagine the size and scale of the tuna they were pulling aboard. Still, even they had a role to play in tagging: early in the 1960s, Frank Mather earned the trust of some of the earliest bluefin purse seiners, and they agreed to allow taggers on board.

According to one fish story documented in Whynott's *Giant Bluefin*, it was a few years after the seiners arrived when a harpooner named Bob Sampson pulled up alongside one to try to nab some big, easy fish. He saw two men in wet suits walking inside the boat's tightly pulled purse seine net, and while he watched, the two men released a big tuna from the net. Sensing easy quarry, Sampson harpooned the fish, and just as quickly one of the men in the net started screaming at him. What he didn't know at the time was that the fish had been released as part of Mather's tagging program. "[Captain] Frank Cyganowski told Bob he could get $5 for returning the tag," writes Whynott. "Bob, not to be outdone, backed up the boat, handed over the tag and asked for his $5—after all, it was a free-swimming fish."

For Amelia's recent ancestors—those Atlantic bluefin tuna swimming off the eastern seaboard throughout the 1960s onward—the region's increase in nautical traffic would have also increased the ocean's natural cacophony. One study by Scripps Institution of Oceanography, based on declassified naval documents, found ocean noise off Southern California increased by tenfold between 1964 and 2004. Italian researchers have found schooling tuna change depth and swimming patterns in reaction to boat noise, so for blue-fin tuna, which can likely hear frequencies between 200 hertz (a foghorn) and 1,000 hertz (a flute), that increase ratcheted up the ambient chaos. They were not only increasingly being pursued over wider distances but also finding it harder to escape humans' clamor.

When he first started catching and tagging bass on Long Island, Al didn't know the term "citizen science" because it hadn't been coined yet. A writer first used the term in a 1989 article in *MIT Technology Review* describing acid rain sample collection by Audubon Society volunteers. The society had long led in that kind of public-facing data collection: in 1900, ornithologist Frank Chapman organized the group's first Christmas bird count, in which 27 volunteers observed 90 different bird species. Mather had already sold Al on the power of carefully collected fishing data, which he had known intuitively ever since he started tagging bass in Lake Ronkonkoma.

As he continued looking for clients who wanted to tag and release fish, Al grew increasingly frustrated by how both striped bass and bluefin populations seemed to be thinning. Every year it seemed harder to find fish for his clients to catch, a trend he had observed ever since the big ships showed up. In the late 1960s the bluefin population had been "ripped apart" by a canning boom on the eastern seaboard, he complained. And massive seiners plundering its waters still netted many times more than his annual catch in a single day. It was a "massacre," he told his friends, one that would indefinitely prevent the bluefin population from recovering.

As Al started focusing on selling his fish-tagging charters to corporate groups, he often found his schedule completely booked up during the busy season, even though he charged as much or more than many other Rhode Island captains. It seemed too good to be true sometimes: catching fish without all the gutting and filleting, plus he got to experience the thrill of convincing other anglers of the joys of releasing a fish instead of killing it. Still, it was impossible to ignore the fact that every year he and his clients seemed to be catching fewer fish. Some days Al complained that the big seiners and draggers, most of which flew under the flags of distant countries, should be blamed for the decline. Those foreign boats should be banned, some raged. The widely held sentiment that foreign boats should be banned—which sometimes manifested as xenophobic rants on regional docks across the United States and Canada—eventually grew into a political movement those countries' governments and others' could no longer deny.

During the Second World War, the United States established a secure zone off the Americas with the goal of preventing Axis-aligned ships from resupplying in South American ports. In the following decades, increasingly powerful and efficient fishing vessels from around the world ranged farther from their home countries and ports as fish stocks crept increasingly offshore. For coastal communities accustomed to an easy day's catch, the idea that foreign ships were taking "their" fish spread like wildfire, including in immigrant-founded ports across the Northeast. It was this widely held belief—that other countries' rapacious fishing habits were to blame for the decline in close-to-home fish stocks—that helped spur the formation of 200-mile "exclusive economic zones" around the world throughout the late 1970s and 1980s.

In 1972, 15 Caribbean nations held a conference to address "problems of the sea," where the concept of a "patrimonial sea" took root. This idea held that each country was entitled to exclusive rights to develop, fish, and conduct research in its own waters, while continuing to allow ships, aircraft, and submarine cables and pipelines unfettered access. A paper submitted in 1971 by the Kenyan government as part of pan-African conversations stated that the "present regime of the high seas benefits only the developed countries," and it defined and made a case for the establishment of exclusive economic zones that granted countries control over their immediate maritime resources—a definition that included fish. As conversations began at the United Nations to codify what would become the 1982 Convention on the Law of the Sea, 7 of 21 detailed sections on resources explicitly addressed what one report called "living resources."

After years of jockeying over details, including one 1974 proposal from Australia that highly migratory fish like bluefin tuna be managed internationally, rough global consensus was reached: coastal states would retain sovereign rights over the waters that abutted their territories, as well as the responsibility to ensure marine populations remained healthy enough to guarantee their continued survival. It was at this moment that a fisheries management tool called maximum sustainable yield (MSY) began creeping into government policy around the world. This concept, which the historians and authors Carmel Finley and Naomi Oreskes have called "a policy disguised as science," eventually allowed the hollowing out of bluefin populations and countless other fish species across our oceans.

To understand how MSY became a cornerstone of global fisheries, one first needs to meet Wilbert M. Chapman. Trained as an ichthyologist at the University of Washington in the 1930s, Chapman spent a year and a half deployed in the eastern Pacific during the Second World War. His job was finding enough food to feed American troops stationed there, and he later returned to the United States passionate about the potential of that region's fish stocks—an opportunity that he said could be as profound as the harvest of buffalo in the American West. In 1948, he was hired as the country's first undersecretary of state for fisheries, a position in which he championed a policy of "maximum production of food from the sea on a sustained basis year after year."

The idea, as he saw it, was to allow individuals and companies to fish until fish stocks declined, at which point restrictions would be put in place. It was a theory based on four assumptions: first, that scientists could accurately estimate fish populations; second, that they could calculate when a maximum sustainable yield was reached; third, that they could then act quickly to implement fishing restrictions; and finally, that they would reopen fisheries only when they could prove populations were healthy. In 1950, Milton C. James, assistant director for the US Fish and Wildlife Service, wryly observed that a fisheries manager arrived armed with "vast, unorganized ignorance, illuminated by occasional flashes of traditional legend, hearsay, inference, assumption, guesswork" and lived a "harassed existence." Even if a manager could protect a fish species by "leaning over backward in regulating [and] giving the resource the benefit of the doubt," there was still the risk of devastating the economic survival of "thousands of individuals, hundreds of communities, and dozens of counties." Fisheries science, it would seem, was an impossible, thankless job with no easy answers.

In 1954, biologist Milner Schaefer proposed a mathematical formula that attempted to describe the logic and science behind MSY. Schaefer, then the director of investigations at the Inter-American Tropical Tuna Commission, built his calculations on population-modeling equations first developed in the 1930s. He named his new model the "Schaefer short-term catch equation," and it, paired with Chapman's theoretical ideas, eventually evolved into today's modern understanding of MSY. Simply, it is the growth of a population of fish plotted against time as an elongated S-shaped growth curve, with the population initially growing quickly as it reproduces, eats, and then reproduces again. As population density increases, food, shelter, and mates become limiting factors, which slows population growth until it levels off at a carrying capacity. What Schaefer codified in 1954 was the idea that if a fish population was kept at half its maximum population size—with the midpoint of the S curve where population growth peaks—then fishers could reliably pull that number of fish from the ocean without wiping out a species completely.

As America entrenched MSY as the keystone of its fisheries management policy and international lobbying efforts, Al and other New England charter skippers continued fishing within this novel and disrupted fisheries regime, trapped between quotas and commercial influence over the fisheries. The unfairness of that differential troubled Al: Why were he and his clients suffering when the fault lay with those entrusted to set sustainable fishing levels for both bluefin tuna and striped bass? "They don't care about the fish that are providing them a living," he told a newspaper reporter. "They don't respect the resource. They don't give anything back."

Many environmental scientists also felt discomfort with what they were seeing. In 1973, Canadian ecologist C. S. Holling published a paper suggesting that MSY "might paradoxically increase the chance for extinctions." Four years later, British Columbia–based fisheries manager Peter Larkin published an article calling for the death of MSY. While the model had once "established order of a sort," he called for a new model with "more sophistication." To conclude, he even wrote a funereal epitaph:

> Here lies the concept, MSY.
> It advocated yields too high,
> And didn't spell out how to slice the pie.
> We bury it with the best of wishes,
> Especially on behalf of fishes.
> We don't know yet what will take its place,
> But hope it's as good for the human race.
> R.I.P.

Similarly discomfited by results of his experiments on the other side of North America, Frank Mather started warning other scientists and fisheries managers about continuing declines in

Atlantic bluefin tuna. He had already seen the devastating effects of the super seining fleet in the mid-1960s and believed his tagging work throughout that decade had played a role in encouraging greater conservation of the species. Still, that didn't keep him from avidly participating in the fishery himself, posing beside giant fish hanging on docks or chartering boats from Bimini to Prince Edward Island to cast his lines.

"I think it's a safe bet to say that if it were not for the tagging program," Mather told Al in 1989, "there would be very little bluefin tuna available today." Mather felt a personal connection to the fishery's origin. He had, after all, been an avid fisherman of bluefin himself and had grown up during an era when the bluefin-catching craze swept across the best anglers from around the world. He also knew how swiftly that frenzy could fundamentally transform a fishery. It was up to him and his colleagues, he felt, to temper the human urge to catch the biggest or most fish until there was nothing left.

CHAPTER THREE

AGE OF GIANTS

~~~~~~~~~~~~~~~

Wedgeport, Way Back When

Why were you ever such a fool as to hook on to one of these
giant creatures in the first place! You must have been mad.
But all fishermen are mad.

—VAN CAMPEN HEILNER, *SALT WATER FISHING*

Over the blue-steel waves off Wedgeport, Nova Scotia,
in 1935, the sun rose slowly and then all at once, first
overtaking the morning stars, then teasing pink on the
clouds, before finally splitting the horizon in two. Michael Lerner,
a 44-year-old heir to a New York City clothing-store fortune, sat
near the boat's bow, gazing at the ocean, as his guides Tommy
Gifford and Lansdell "Bounce" Anderson chapped their hands on
the oars. For hours already, their boat had bobbed fruitlessly
without a single bite on Lerner's bait. Finally the rod twanged
with the hit; it was a brutal, fast snatch. His quarry, at last.

The physics of Lerner's fight with the giant Atlantic bluefin tuna were simple: the tuna was 300 pounds of muscle hooked to a line made of 54 braided strands of spun linen. That line threaded through a rod held by Lerner, who had been strapped to a boat-mounted swiveling chair that prevented him from being pulled headfirst into the ocean. Catching a bluefin tuna on rod and reel required the skill, strength, and endurance of a world-class fisherman, and every piece of gear had to work—from the bamboo rod to its arched metal hook. And as of that day in Wedgeport, it had never been done before.

Deep underwater, the hooked bluefin followed instinct, kicking its powerful sickled tail as it rocketed away from the dory. Like a speeding car, it ramped up its speed, drawing on the digested caloric power of all the tiny fish it had eaten that week on the marine bank locals had dubbed Soldier's Rip, a bountiful patch of ocean about 15 kilometers or so offshore from the village. That power fed its warm organs and dense red blood, its thick muscles throbbing with lactic acid as it pulled and ran. The fish's pectoral fins slotted into its sides as it strained against Lerner's rod, its skin flashing a rainbow of colors in agitation.

The metal mechanism of Lerner's reel screamed as it spun, letting out line faster than his eyes could follow. Even still, the fish towed the wooden dory across the glittering chop. Lerner fought to keep line on the reel without breaking the tenuous connection. He knew it would be something like this, the world crystallized around his human body in a single second: water, wind, and sun; man and fish. But this wasn't a gentle tease from the ocean's depths. This was a tug-of-war with a bear. When Ernest Hemingway, a fishing friend of Lerner's, first saw a big tuna off the coast of Spain, he was shocked at how the giant fish leapt clear of the water, falling back against it "with a noise like horses jumping off a dock." Anyone capable and canny enough to catch a fish that size, Hemingway wrote with awe, could "enter unabashed into the presence of the very elder gods."

After being dragged around the ocean for nearly half an hour, Lerner started to tire. But so, too, had the fish. With a final, deep tug, Lerner pulled the bluefin's gleaming, torpedo-shaped body alongside the boat, one smooth side of shimmering skin tipped toward the sky. A golf-ball-sized eye gleamed in its blue-black head, as its sharp pectoral fin slapped the air fruitlessly. The fish, already close to death, flapped its fins with exhaustion, yet it still took every sickled gaff and ounce of strength the three men had to pull its bulk over the dory's gunwale. No sooner had Gifford baited and cast the next hook than another tuna, this one even larger, was on the line. Within another hour, Lerner had landed this fish too, also more than 300 pounds, before they called it a day. The trio headed back to port, Gifford and the mate pulling the boat's oars with bleeding hands and aching backs.

DURING AN ERA when commercial fishing and adventure-seeking tourism started to boom and converge, bluefin tuna transformed Wedgeport's fortunes. Under its cold waters, giant bluefin tuna schooled at the turbulent waters where two prevailing currents collided. From the surface, the rip appeared as a flat plate of ocean ringed by curling waves that seemed to come out of nowhere, and for decades Nova Scotians had witnessed schools of huge tuna congregating there. Punctuated by a massive undersea bank that pushed nutrients and animal life upward toward the surface, tuna grew huge on that rip, plump from gorging on schools of herring, mackerel, and squid. Catching fish was easy on the rip, and catching fish was what had brought British and French colonizers to Nova Scotia in the first place.

Only a few days before he arrived in Wedgeport in 1935, Lerner had already given up on his dreams of catching a giant bluefin. The avid sportsman and angler had traveled to the remote southern coast of Nova Scotia to join paid guide Gifford for a week of fishing. But all the money in the world can't conjure a fish that doesn't bite, and they hadn't landed a single noteworthy fish. Frustrated and disappointed, the pair decided to head west by train and catch the next ferry across the Gulf of Maine back home to the United States. As their steam-powered train chugged along, Lerner couldn't bear its painfully slow speed. After long minutes of complaining, the pair lunged off their train car at the train's next stop, lugging bags, hats, and all, and flagged down a rickety passing car that, to Gifford's eye, looked like "one of the first automobiles ever made." They negotiated a fare, loaded their tackle, and bounced off heading west down the winding gravel road.

Driving along past wind-whipped pines and glacier-hewn boulders, the men eventually stopped for gas at a tiny, ramshackle stand. Inside, pinned to a plank wall, they spied a photo that beggared belief: a black-and-white newspaper clipping of a tuna larger than a boulder. At the bottom of the paper's curling edges someone had scrawled "1,100 pounds." Sensing possibility, the Americans interrogated the man behind the counter. That fish?

Sure, he said in the region's lilting Acadian accent. That fish had been caught in Wedgeport, a fishing town to the southeast named for its triangular wedge of land that hangs into the chilly Atlantic like a lonely, thick icicle. That fish inspired the pair to give bluefin fishing in Nova Scotia one more shot.

Late that afternoon, Lerner and Gifford sputtered into Wedgeport dusty from their journey and headed down a slope into the town's port, asking around for someone who could help them catch a tuna. In a town where women sold hooked rugs from the side of the road for cash and most men held down more than a few jobs, the money those American anglers threw around gleamed. After their first few queries, Lerner found Evée LeBlanc.

LeBlanc had ancestors who had been part of the region's early wave of colonial arrivals from France who, between 1755 and 1763, were forcibly deported from Nova Scotia to the United States by the British. But their connection to the region remained strong, and by the early 1800s, Acadians had resettled the western end of Nova Scotia's long horizontal coastline in droves. They traveled north from New York, Connecticut, Massachusetts, South Carolina, and Maryland, where they had lived since being unceremoniously stranded, many brutally separated from husbands, wives, and children who had been intentionally put on different ships. Upon returning to their newly declared homeland in Canada, the Acadians built tight-knit social groups and harbored a fierce protectiveness of their Francophone roots and way of life. That, to a profound degree, included sailing, boatbuilding, and catching fish.

In the 1920s, LeBlanc started harpooning the bluefin tuna

alongside two other Wedgeport-based fishermen. "How the men hated those tuna, those horse mackerel!" David MacDonald wrote in 1955 in Canada's *Maclean's* magazine. "As big as a thousand pounds, they wrecked nets. Speared, they fought for hours. And all the monsters were worth was a mean three cents per pound at canneries along the shore." By the early 1930s, LeBlanc had already repeatedly tried to catch giant bluefin on rod and reel, a piece of fishing equipment designed for smaller fish. Instead of a delicate linen line, he rigged his rod with a double steel line tied to piano wire. Before the invention of fiberglass rods, a hard tuna strike could reduce a fisherman's bamboo rod to shards. At the time, less sportingly but more lucratively, Wedgeport's fishermen also corralled bluefin in nets in the open ocean en masse. Once the bluefin were netted, the fisherman pulled them to shore, dragging the ponderous catch behind their boats and eventually aground. Helpless in the shallow water, the fish were killed and sent to either Boston's fish market or a cannery. In 1932, Nova Scotia netters had landed 204 tonnes of tuna, nearly double the previous year's catch. It was a fishery for flesh and sustenance, not an activity fit for a gentleman.

Despite having no luck with rod and reel, LeBlanc's brother Louis and another friend did manage to harpoon the largest bluefin tuna ever landed in Wedgeport in 1934, a 1,100-pound giant they stretched out at the wharf for gawkers. That was likely the fish Lerner eventually glimpsed in the photo at the gas station—a bluefin that had bumped history off its steady trajectory as a low-value novelty fish. That picture set the hook. The circle complete, the two Americans rented LeBlanc's small wooden fishing boat, or dory, and rigged a swivel to the chair on its front—a big-fish fighting trick Lerner had picked up in the Caribbean—and hired a third man to pull the boat's second oar. The next day they left at dawn and returned home with their two giant fish: the first rod-and-reel tuna ever landed in the region.

By the time Lerner and Gifford arrived back at the dock, Wedgeport boys were shouting news of the fish on the streets, and the men were passed around for backslaps and handshakes. The next day, the pair headed out for more fishing, only to discover a port packed with boats, each crammed with looky-loos who wanted to see a bluefin caught by rod and reel for themselves. The flotilla, including one boat with a brass band aboard, headed toward the bank where tuna swam; Lerner caught his next two giant tuna to the melodic notes of horns floating across the waves.

Energized by his success out on the water, Lerner excitedly summoned his friend, journalist and angler Kip Farrington Jr., who was fishing nearby, to join in on the action. Farrington, who was traveling and fishing with his wife, Sara Chisholm, couldn't pass up the invitation. By the end of the couple's first day chasing tuna in Wedgeport, both had caught their own fish, making Chisholm the first woman on record to catch a tuna on rod and reel.

After his Nova Scotia trip ended and he returned to New York, Lerner passed along photos of the massive fish to some sportswriters he knew, who published them in newspapers and magazines. Soon the international wires to Yarmouth were humming with interest. Lerner, who, according to Gifford, "would have severed a leg or arm as readily as he would the line if there was a fish on the other end of it," returned to Wedgeport within weeks. By the end of his second trip, Lerner had landed 26 giant bluefin tuna.

As the news of Lerner's success catching bluefin in Wedgeport spread, the world's fishing illuminati responded. Amelia Earhart, by then widely known for her aerial feats, spent two weeks traveling and tuna fishing with friends along Nova Scotia's coastline. In 1935, coal heir and sportsman Van Campen Heilner included a chapter on giant bluefin fishing in Wedgeport in his definitive *Salt Water Fishing*. In a vivid color plate printed within the book entitled "Nova Scotia Sleigh Ride," a giant fish pulls a two-man dory against a salmon-colored sky. "The tuna is there now, his great blue-black back just beneath the surface," Heilner wrote. "He's so big he scares you. It can't be possible that you've almost mastered him. . . . There is a wild roar of waters as the great fish threshes from side to side and is drawn quivering over the stern. The giant mackerel is yours."

From those earliest days, Lerner saw the possibilities for Wedgeport. He called the town's top fishermen into a meeting and laid it out for them. After Lerner's presentation, it was clear the American felt their long-maligned bluefin could draw anglers "from the four corners of the world," according to Wedgeport fisherman and former army captain Israel Pothier, although he personally saw Lerner as "batty or a teller of tall tales." But Lerner told them to clean their boats, install fishing chairs and toilets, buy their own bluefin fishing tackle, and get ready for the world's arrival.

Wedgeport's citizens didn't waste any time throwing themselves into the enterprise, opening hotels and restaurants, printing business cards, and placing newspaper ads. The hype even reached the Oval Office: one day in 1936, the year of his reelection, US president Franklin Delano Roosevelt fished for tuna on Soldier's Rip (even all the power in the world can't guarantee a fish on the line; he didn't get a nibble). Babe Ruth also visited that year.

As wealthy American anglers scrambled northward, Farrington had an idea: they would recruit teams of top fishermen from around the world and invite them to Wedgeport for a few days of fishing. It would be an international tournament featuring giant bluefin with a wide silver trophy, a prize donated by Boston's Eastern Steamship Lines president Alton Sharp.

Starting in 1937, teams from 28 countries, along with writers, photographers, and gawkers, eventually overran the town and its population of 1,000 people for the annual event. Homes were converted to hotels, housewives volunteered as chefs, and small groups of locals banded together to form crews. Paid for by the province of Nova Scotia's trade board, each crew consisted of a captain, a guide familiar with tuna locations and behavior, and a chummer, whose sole job was pulling herring out of a barrel, cutting them into pieces, and throwing them overboard to draw the tuna close. Each member of a fishing team was given a time slot, and the instant a fish struck the bait, it was that fisherman's turn on the rod. In addition to daily pay, crew members were often given the landed tunas to sell, and the few sports who took the fish back to their home countries for stuffing and mounting often paid the locals for the market value of each fish. At the end of the tournament, once they had landed their quarry, watched organizers hand out prizes, and swilled champagne from the silver trophy, most anglers lost interest in the tunas' heavy carcasses. The fish hung on the docks, their skin drying and cracking in the ocean breeze until they were cut down and either sold to a cannery or dumped.

~

Often uncomfortably, the growth in Wedgeport's tourism economy posed a double-edged sword. Marketed abroad by the province's government as a sportsmen's paradise, Wedgeport had locals who were pegged as quaint, antimodern "folk," the inhabitants of and guides to an unspoiled wilderness of beauty and bounty—at least for those willing to pay for the privilege. And pay they did. By the mid-1900s, one report estimated that anglers annually spent at least $2 million in today's dollars in the town, with local fishermen making nearly twice as much as their counterparts elsewhere in the province. That money "bought cars, launched boats, sent sons off to college," recounted *Maclean's,* and helped build new houses and fix up ramshackle ones.

The festive spirit of the annual tournament, according to another magazine account, provided part of its allure. Locals sipped martinis with Brazilian coffee kings while the Cubans tried to teach everyone how to samba. "There are tournaments with more contestants, but they're all so parochial," said Toronto's Tom Wheeler, an organizer of the British Empire team. "Any angler would give his eyeteeth to compete at Wedgeport." One year, according to the *Maclean's* report, a locomotive engineer traveled 38,000 kilometers from South Africa, using up three years of vacation time and his life savings to join Wheeler's squad—but he didn't end up catching a single fish. Soon after arriving for the 1939 tournament, teams were disappointed, but not shocked, to receive news that the Second World War had begun. "The French team said, 'Nous allons chez nous tout de suite. . . . Goodbye, Bonjour,'" recalled Israel Pothier.

Organizers canceled that match and subsequent matches for the duration of the war, but the event resumed to great fanfare in

1946. That year, Farrington wrote about his return to the town for *Maclean's*. "Nowhere had I fished where so many big fish were surfacing at one time," he wrote. "I don't believe there was a fish showing that weighed less than 500 pounds." He watched with amazement as hundreds of fish jumped, rolled, and swam at high speeds all around his boat at the rip. Tourists returned in droves, and the town breathed a sigh of relief as their wallets and booking calendars filled up once again.

For subsequent decades, a visit to Wedgeport remained a bucket-list item for any high-class fisherman in the know, just as a boom of postwar prosperity and tourism brought with it new opportunities for towns trying to draw new visitors. And one scrappy Canadian widow played an essential role in bringing the glamour and excitement of Nova Scotia's bluefin industry to a wider global audience.

In 1946—the year Wedgeport's tuna tournament resumed after the war—Margaret Perry, a 41-year-old, curly-haired widow, arrived in Wedgeport lugging a suitcase-sized 16 mm film camera and her heavy, awkward tripod. Recently hired as the only film-making employee of Nova Scotia's industry and publicity department, Perry had come to Wedgeport, like so many others before her, to witness the bluefin spectacle. Yet her aim was different: she wanted to curry the story of the spectacle to greater fame world-wide and help draw tourists to her home region. She had a front-row seat to the changes that a fishing boom could bring to a down-on-its-luck fishing village and was aiming to capture, in her film, how entangled a community could become with a fish on which its fortunes depended.

For weeks, Perry set up shop in the village, becoming a familiar sight on docks and boats in her pleated khaki pants and hair that blew wildly in the wind as she framed shots through her camera's viewfinder. She talked to fishermen who told her about their town and way of life, making notes in a notebook she juggled along with all her other equipment. She captured shots of the town's church, schoolhouse, and cooperative store, of yoked oxen pulling carts, cows browsing on beaches, and men beside them digging clams with hoes and buckets. Out on the water, she shot big tuna surfacing in a rise of water and bubbles, lured with fish scraps and pursued with long harpoons from the prows of dories. Aboard a rocking boat, she filmed the delicate, gory procedure of removing a bait herring's backbone, so it would wiggle more realistically in the water on the line.

To demonstrate migration patterns of bluefin, as they were known at the time, Perry built an elaborate lightbox of projected, cutout shapes of continents bordering the Atlantic Ocean. With her camera running, Perry used a magnet to move tiny white paper tuna, filled with metal and drawn across a projector screen's blue backdrop. In the film, the tiny white torpedoes make their way down the coast of Europe, eventually striking the Gulf Stream off Africa's western coast. The current then draws them toward a group of islands labeled "Indies," where, according to the voice-over, the fish stopped to breed before making their way up toward Canada's east coast.

Perry delighted at the participation of women in the sport and wrote a section of script for the film that runs over a jaunty instrumental interlude. "Women may be considered the weaker sex, but watch Georgia Manning, who has many record catches to her credit," enthuses the film's narrator near the film's end.

"She has a strike already!" In the vignette, Manning, a brown-haired angler in rainproof pants and white gloves, rocks back and forth in the straps of the fighting chair's cloth harness as she smiles widely. Together, Georgia and her husband, John—the "Angling Mannings"—landed 43 tuna weighing a combined 12.5 tonnes between 1938 and 1946. "No, tuna fishing is not just a man's game," the voice-over declares.

Privately, Perry marveled at the road she had traveled that had brought her to directing her very first documentary. A decade earlier her life had been mostly unremarkable. She was born Margaret Rice just after the turn of the century in the small village of Upper Mills, New Brunswick, on the cusp of Canada's border with the wilds of eastern Maine. She trained as a stenographer, one of the few careers welcoming to women at the time, but hated that job and being stuck within an office's four walls every day. Her real home was the outdoors, the deep forests and rocky beaches that crept along the Atlantic coastline.

In 1934, Margaret Rice married Stanley C. Perry, a geology and minerology professor at the University of New Brunswick, and they settled into a happy life together. Two years later, Stanley died suddenly when his car collided with a speeding fire truck. He left his wife behind, pregnant and with no income. A reeling Perry leaned on her growing passion for photography and started taking photos and writing for magazines. She signed up for a correspondence course in "still and movie photography," and when her son was born, she named him Stan after his late father.

As Perry poured herself into learning the technical and artistic craft of making films, she worked as a projectionist for the newly formed National Film Board of Canada (NFB) during the Second World War. She lugged projection gear and delicate reels to community centers and church basements across the Maritimes, providing those communities with a glimpse of valor and international warfare beyond their plowed fields and dusty gravel roads. In her spare time, Perry made 10-minute documentary films about life in Miramichi, a rural northeastern region of New Brunswick. The amateur shorts caught the attention of a local member of Parliament who recommended the nascent filmmaker to NFB founder John Grierson, the man who first coined the word "documentary" in 1926.

In June 1942, when her son, Stan, was only five years old, Perry arrived in Ottawa, Canada's capital, where she wrote scripts and directed films, including one on the cooperative fishing movement in the Bay of Fundy and another on Nova Scotia's lobster fishery. When the war ended, that province's government offered her a job in its tourism department, where she became its first director of motion picture publicity films.

For nearly 15 years, Perry worked as the sole filmmaker and staff member of the Nova Scotia Film Bureau, where she wore all the hats: cinematographer, director, editor, location scout, scriptwriter, sound recorder, and driver. "Sometimes I didn't know if I was going to sink or swim," she recalled to writer Olga Denisko in 1975. "The only good thing about it was that nobody else knew more than I did, and so it was just up to me. And I made out." In her work, Perry was drawn to the dramatic tensions threaded throughout Atlantic Canada's postwar society, and Wedgeport's booming bluefin industry provided a tantalizing lacuna through which to tell that story. Her job, after all, straddled the worlds of promoting both tourism and industry, which, according to one critic, created a consistent tension within her films "between images evoking an anti-modern culture amidst natural splendor, and those that glorify relentless industrial progress." Looking at a photo of Perry standing on a dock, I found myself struck by a deep affinity for her and her work: a mother, trying to conceive of a largely male sport and profession, while still claiming it as her own, to bear witness and to tell its story.

The film that brought her to Wedgeport, *Battling "Blue-Fins,"* was her first major project in this role and won Perry an award in 1951 from a documentary film festival in Rome. It cast a bright and dazzlingly public light on a niche industry that had, up to that point, mostly grown through word of mouth. Ultimately, the film showed Wedgeport's townspeople and wealthy tourists engaged in a delicate dance, one rooted in the act of catching tuna but embellished by an exotic experience of each other. It captured a region in the throes of transformation, as it made awkward, delightful changes to adapt to modernity.

In the postwar landscape, Wedgeport was a sportsman's aspiration. And with so much money and machismo flowing through the town, it was only a matter of time before Wedgeport experienced its first international incident, which arrived rather bizarrely from the reaches of Outer Baldonia—one of the world's first modern micronations and, as I was to discover, the next target, after Margaret Perry, of my bluefin tuna fixation.

IN 1948, RUSSELL Arundel first spotted the rocky bluffs of Outer Bald Tusket Island, a four-acre island five nautical miles off Nova Scotia's coast, near Wedgeport. The island, a long, wedge-shaped slice of green scrub and white stone, lay near the fishing grounds where tuna congregated and had historically been used as a stopping-over point for fishermen. Arundel, a tuna angler and chairman of the Pepsi-Cola Bottling Company based in Washington, DC, wanted to buy it, and he was a man who got what he wanted. Within a year of making the purchase, which cost him $750 (about $9,000 today), Arundel declared himself the Prince of Princes of Outer Baldonia. "Back in Washington, the deed in my pocket and a drink in my hand, the Principality of Outer Baldonia began to take shape," he told *Esquire* in 1953.

Known as a hard partier on Capitol Hill, Arundel issued his own Declaration of Independence, which, in addition to banning women outright, declared fishermen to be "a race alone." He endowed them with the following inalienable rights: "The right of freedom from question, nagging, shaving, interruption, women, taxes, politics, wars, monologues, cant and inhibition. . . . The right to swear, lie, drink, gamble. . . . The right to sleep all day and stay up all night." The group's responsibilities included fishing as well as the exportation of empty rum and beer bottles.

Arundel appointed DC tax attorney Prew Savoy as prince regent, minister of state, and "Ambassador to the United States without Portfolio or Credentials" and offered 20 wealthy white Americans and Nova Scotians princehoods. Citizens of the "nation" included 70 members of the Wedgeport Tuna Guides Association—they were declared "eight-star admirals" in the Outer Baldonian navy. The principality's cloth flag was a white circle with a Dutch-blue tuna tail on a green background. Arundel minted the kingdom's currency, the tunar, in $1 coins and $25,000 bills, the latter of which were printed with the slogan "War on Poverty." Alongside his friend, the watchmaker Arde Bulova, he even designed an 18-karat gold pendant watch emblazoned with the Outer Baldonia seal, which he gifted to family and close friends. Despite hiring a group of locals to build a small, one-room cottage using smooth beach stones, Arundel spent only a single night on the island, declaring it inhospitable due to mosquitos and cold, gusting winds. But that didn't stop him from listing the nation in the Washington, DC, phonebook or compelling a friend at the Rand McNally publishing house to include his island on a published map.

In 1952, in the early days of the Cold War, it appeared as if some Soviets didn't get the joke. Writing in Moscow's *Literaturnaya Gazeta*, a letter writer named "L. Chernaya" accused Arundel of granting his subjects the unrestricted right "to tell lies, to be rude." Arundel's rule of Outer Baldonia, Chernaya continued, represented the worst of America's imperialism and the country's export of capitalism worldwide. (One writer at the time did suggest the letter may have been Russian-facing mockery.) In response, Arundel wrote to Russian leaders threatening to break off diplomatic relations, letters that drew snickering public support from the Halifax Yacht Club and Wedgeport fishermen. International newspapers and magazines soon caught wind of the controversy, and in 1967, Nova Scotia's legislature officially recognized the micronation, stipulating that its citizens must nevertheless continue to pay provincial taxes. In 1968, Arundel told a reporter that he never heard back from the Soviets. "And that is how I won the war against Russia," he declared.

Ultimately, Arundel's fidelity to his own nation proved fickle when, starting in the 1950s, Wedgeport's bluefin started disappearing. For reasons scientists and fishermen couldn't explain, the huge tuna that had once prowled Soldier's Rip simply stopped showing up. Some fishermen blamed the Japanese longliners that had recently arrived, while others hypothesized natural changes in ocean currents and food sources had drawn the fish elsewhere. Others attributed the slump to a local priest who had threatened his parishioners that if they didn't stop tuna fishing on Sundays instead of attending church that the fish would go away. In either case, in 1957, the International Tuna Tournament shifted to the community of St. Mary's Bay, and in 1976 was canceled altogether. Arundel, who donated his island to a regional nature conservancy, stopped making the trip to Wedgeport. A quarter century later, one Canadian government report stated that the western Atlantic tuna population of that era had been "seriously depleted, largely because of severe fishing pressure."

Losing myself in Perry and Arundel's paired projects, both clad in the trappings of a trivial romp, gave me a strange and uncomfortable frisson: of how quickly luck struck Wedgeport and how quickly that fortune faded. Gone were the martinis and the felt hats. These days in Wedgeport, it's hard to find a good restaurant open late on a weeknight, and my off-season visit to the town's tuna museum and café required someone calling in a favor to let me into the dark, shuttered building. Inside, I marveled at the giant rods, at the stuffed fish, and at a black-and-white portrait of an Indigenous Mi'kmaq woman in full regalia standing beside a giant fish she had caught. Hers is one of so many stories, barely remembered except by old-timers nursing bitter coffees at the bar.

In a region founded by farmers and fishermen, many accus-

tomed to scraping a subsistence living from the land and sea, the
arrival of new money, tourism, and a powerful globalized fishing
economy resulted in a capitalist experiment. Those huge bluefin,
found nowhere else in the world, rapidly transformed Wedgeport
into a major tourist town, and the money and opportunity flowing from those fish transformed lives across the region. Yet trophy fishing is, at its heart, an exploitative endeavor, and as the
object of a fancy, exclusive sport championed by the era's industrial elites—for whom opportunism was a way of life—the town
had made a calculated risk.

With every huge fish landed on Wedgeport's docks or left to die
hanging from a giant beam in the sun, groups of anglers posing
in front of its body, the Atlantic bluefin population lost some of its
most powerful, most prodigious breeding adults. Fishermen gave
those fish away, threw them away, or sold them for pennies. They
were a commodity for status, for recreation, and a way for families
to send their children away to school to become doctors or lawyers. But as Wedgeport discovered, an economy based primarily
on a natural resource can be a tenuous thing, particularly if that
natural resource has a threshold that no man or woman can
easily see.

CHAPTER FOUR

BEFORE THE STORM

~~~~~~~~~~

AI, 1970s

GILBERT: A dreamer is one who can only find his
way by moonlight, and his punishment is that he sees
the dawn before the rest of the world.
ERNEST: His punishment?
GILBERT: And his reward.

—OSCAR WILDE, "THE CRITIC AS ARTIST"

The first tuna—Amelia's earliest ancestor—opened its eyes onto an ocean riven by change. It was the late Paleocene, between 65 and 55 million years ago, the period following the dinosaurs' extinction but before the Himalayas formed, when volcanoes still regularly thrust upward from the earth's cooling mantle. Bony fishes were evolving from the same primordial soup that birthed sea urchins and seals and eventually subdivided into the Scombridae family, which included tuna. Tectonic changes during the era then disrupted the tropical Tethys Sea, a warm body of water that had previously divided the world's two super-continents and would eventually become the Mediterranean. This kick-started a period of rapid ocean cooling that forced bluefin to travel farther afield for prey and likely resulted in the fish's extra-ordinary endothermic evolution, which eventually allowed it to warm its muscles, organs, and brain using what is essentially a built-in heat pump.

With the ability to travel huge distances unlocked, these new bluefin spread widely and boomed, branching into *Thunnus thynnus* in the Atlantic Ocean, *Thunnus orientalis* in the Pacific Ocean, and *Thunnus maccoyii* in the open waters of the southern hemi-sphere. The Atlantic bluefin, not forgetting the warm waters of its own birth, sought out those same waters when it came time to spawn, a voyage that necessitated huge journeys, which it made using its refined senses for salinity, currents, and possibly even electromagnetism from trace metal deposits in its skull. Those ancient underwater trails eventually led it to the Mediterranean in the east and to the Gulf of Mexico in the west, two watery patches reminiscent of its earliest home waters. When Al Anderson daydreamed, he liked to believe that within every bluefin tuna rested that ageless genetic memory, an encoding of knowledge, passed down fish to fish, that resulted in the strong, wily bluefin he fished from Rhode Island's waters.

ON THANKSGIVING WEEKEND in 1970, a time of year when he'd otherwise be out tagging, Al packed up to spend a few nights at his parents' New Jersey home on Safran Avenue. Arriving off the busy highway, he pulled onto their quiet, tree-lined street, hoisted his bag, and spent the evening under the roof he had helped build. He slept late as much as he could and helped his mother fix a few things around the house, but soon his orderly life, once again, tilted off its axis. On that Thanksgiving Saturday night, 60-year-old Arthur plummeted to the living room floor, dying of a heart attack as Isabelle and Al crouched helpless beside him.

At the time, Al taught biology at Westerly High School while renting a modest apartment, and had been working hard to establish himself as a respected fisherman and generally science-minded fellow in the Narragansett community. In his first few years teaching at the school, he started a co-ed student fishing club that allowed him to pass on his growing knowledge of the waters around Block Island. A few times a year, he and his students would ride school buses to the nearby Point Judith Harbor of Refuge, where they'd climb aboard one of the marina's many party boats or commercial fishing vessels. He marveled at the transformation in the young teens as they became anglers, proudly holding dripping codfish up to the camera with crooked smiles. In the evenings, if he wasn't fishing, he'd be fixing gear or filing the paperwork needed to maintain his US Coast Guard charter skipper's license. During his early tagging years he operated the boat he bought in college, that leaky four-meter plywood skiff he had to bail out every 20 minutes or it would sink. (He knew it was time to get the bucket when his tackle box began to float.) By the time Arthur died, Al had saved up and graduated to a six-meter Aquasport he bought with cash.

Losing his father changed something in Al: his mother sold the house on Safran Avenue for another home in Rhode Island and offered Al the chance to live there with her. In those days, he accepted—but couldn't quite come to terms with—the idea that teaching and fishing and living with his mother would be the extent of his life. He was still young and handsome, and he wanted someone to share his life with, a smart, savvy woman who could accept his passion for fishing and put up with, as he saw them, his "quirks."

By the early 1970s, Al also started to realize that his bluefin fishing hobby could make him some serious money. That decade, new buyers of bluefin began popping up around New England ports, including in Rhode Island. They were foreigners who had never been in the area before, mostly hailing from Japan. These buyers often paid previously unheard-of prices for top-quality bluefin, although sometimes their demands for how the fish were to be stored and processed seemed frustratingly unreasonable to the local "salts," a dockside nickname for longtime fishermen.

It wasn't widely known outside fishing circles at the time, but overfishing of Japanese Pacific bluefin stocks had suddenly crashed that country's bluefin populations, and demand for alternative sources of the fish had begun to spiderweb across the world. Between 1962 and 1967, Japanese boats caught between 5,000 and 12,000 tonnes of bluefin off the coast of Brazil; soon after, the fish virtually disappeared from Brazilian waters. A similar crash in the North Sea and the Norwegian Sea occurred in 1963.

In the 1960s, laws concerning where international fleets could legally fish were still evolving. After the Second World War, Chile, Ecuador, and Peru were the first countries to establish a

"200-mile limit," a declaration that underwater territory would be considered off-limits to deep-sea fishing and whaling by other countries' fishing fleets. The United States and Canada were slower to claim their own limits, and starting in 1963, the fleets of "super seiners," as Mather referred to them, began showing up.

Responding to drops in stocks caused by heavy fishing pressure across the world for species including bluefin, the Brazilian government hosted a conference in Rio de Janeiro three years later, in 1966. There, 17 countries, including the United States, signed on to an agreement establishing the International Commission for the Conservation of Atlantic Tunas (ICCAT). By 1971, the agency, which pulled its structure roughly from the United Nations model and was founded under the authority of the UN's Food and Agriculture Organization, had moved its headquarters to Spain—a choice attributed to the country's high tuna catches and lobbied for by the country's dictator Francisco Franco.

Within a few months, ICCAT administrators were setting up international meetings at which member countries could raise concerns about tuna fishing levels and management, and where they would ultimately divide the spoils of wide-ranging oceanic species including bluefin. Yet it wasn't the length or location of a member state's coastline that determined how tuna catches were divided. Instead, economically developed countries like Canada and the United States used historical catch data to stake their claims on future catches—something smaller countries protested but were largely helpless to stop. For many, ICCAT represented a wolf in secretarial glasses, mandated to carefully record in its ledgers how many sheep were to be consumed, and by whom.

Around the same time, Frank Mather's tagging experiments using data from Al and other taggers were sending him clear signals that bluefin tuna were being overfished. It seemed obvious to Mather that the species would be in danger if catch limits weren't immediately enacted, and he thought he could prove it. His data showed that rates of tagged fish being re-caught were extremely high, which suggested to him that changes in the fish's population structure could be directly attributed to overfishing. Mather, who worried about the fish he loved to catch, called the situation "catastrophic." During the same period, he remained active in the international scientific community's inquiries into bluefin, including sitting on the FAO's Continuing Committee for the Facilitation of Bluefin Tuna Research, corresponding with bluefin researchers around the world.

As early as 1970, many policymakers and scholars knew that the planet could not sustain that era's current extraction practices and levels—paired with ongoing human pollution and ecosystem destruction—for any meaningful period of time. They bristled at the degree and side effects of extraction, an abuse of nature on a global scale in the name of progress, expedience, and profit. "[We] will have to tackle on a hitherto unprecedented scale the thorny task of regulating industrialized fishing in international waters," the academic George F. Kennan wrote in a 1970 *Foreign Affairs* essay entitled "To Prevent a World Wasteland." In the piece, he suggests that a global fleet of independent patrol vessels with enforcement power might be the only solution to the race to the bottom. "Exploitative motives cannot usefully be mingled with conservational ones," he wrote. "The principle should be that one exploits what a careful regard for the needs of conservation leaves to be exploited, not that one conserves what a liberal indulgence of the impulse to development leaves to be conserved."

While the US Department of Commerce offered subsidies to commercial bluefin fishermen, Mather spent his own money traveling to Washington, DC, where he tried to convince lawmakers that the bluefin's late spawning age, which he had concluded hovered around five years old, meant it reproduced slowly and made it vulnerable to overfishing. He also warned them that even the sparse data points he had gathered were incomplete: in 1974, he noted that the volume of transatlantic migrations of fish from the eastern (European) stock to the western (North American) stock could scuttle any certainty scientists had that their stock assessments were correct. By 1966, Mather had documented more than 14 tagged fish—caught and tagged at two years old off the US coast in 1954—that were eventually caught in the Bay of Biscay, just off France's arcing coastline.

Those recoveries echoed stories Mather found, dating back to the early 1900s, of fishermen in the western Atlantic discovering Mediterranean fishing hooks lodged in the mouths of bluefin. While those numbers of ocean-crossing fish paled in comparison with the huge number of bluefin tuna in the Atlantic, Mather still strongly believed it was a phenomenon that should give fisheries managers pause. If more bluefin than they knew regularly crossed the Atlantic, the predictive math they used to ensure the species' long-term health couldn't be trusted.

Why exactly these fish were crossing the Atlantic remained a mystery and challenged the prevailing theory that Atlantic bluefin lived and spawned in two independent groups. But despite Mather's data, federal fisheries managers were unswayed, telling him that he didn't, after all, have a degree in biology and should leave the number-crunching to the professionals. Still, those transatlantic recaptures gave Mather pause. If any progress was to be made on understanding and protecting bluefin, he believed, it would have to be the result of sustained international effort. "This species cannot be studied successfully on a piecemeal or limited area basis," he wrote in one unpublished report from the late 1970s.[1]

As bluefins' transatlantic voyages defied international borders, their spiking value and enormous range made them an easy target for growing global fishing fleets. At ICCAT meetings in 1973 and 1974, the US delegation put forward a proposal that catches of large bluefin on either side of the Atlantic be chopped by 25 percent from the average catch levels in 1968 to 1971, with a 50 percent catch reduction in fish seven years and younger and a minimum catch size of 14 pounds, but the proposal failed. In 1975, however, delegates did agree to a massively scaled-back version of the plan, agreeing to the size limit and a voluntary quota—although this latter condition was more of a suggestion, without any enforcement or detailed catch numbers attached.

Roger Hillhouse, a tuna spotter pilot and purse seiner owner, agreed with Mather's dire predictions—one of the few commercial fishermen to do so publicly at the time. In 1974, Hillhouse and another captain showed up at an ICCAT meeting in Madrid with a request that his industry be further regulated to help preserve bluefin populations; he suggested a size limit of 15 to 20 pounds, which the organization rejected. The following season, Hillhouse and two other captains set a voluntary per-harvester quota of 998 tonnes, and Hillhouse kept his word: for the next decade, he docked his three purse seiners after they reached that threshold. He also wrote a public letter blasting his industry for wiping out bluefin alongside the passenger pigeon and buffalo. "His colleagues beat him down enough that he kind of kept quiet after that," Mather once said in a book interview. "He's the only commercial guy I knew who had a farsighted attitude."

In August 1975, in part due to pressure exerted by Mather, US president Gerald Ford signed the Atlantic Tunas Convention Act of 1975, which made ICCAT regulations enforceable within the country. Debate over the legislation had been intense, with many commercial fishermen fighting any future prospect of catch quotas or limits that the organization might potentially impose. Mather, not accustomed to a public spotlight, struggled with his new role as the target of many fishermen's ire. But an op-ed published in *Right Rigger* magazine after the bill's signing celebrated Mather and the importance of the legislation, which was the country's first-ever substantial conservation plan for oceanic gamefish. "Conservation programs are never popular, especially when they are needed," it noted.

The following year, the United States passed the Magnuson Fishery Conservation and Management Act, which put federal management of the nation's fisheries under the National Marine Fisheries Service and eight regional councils.[2] The move built a framework for protecting species and restricting foreign fishing fleets from entering the country's exclusive economic zone within 200 miles of its coastlines. And from the very beginning of those regional councils, it was clear that commercial fisheries and political interests would set the agenda. Environmentalists and watchdog groups at the meetings were few and far between, and delegates were focused on splitting marine spoils between states, catching foreign vessels infringing on US waters, and promoting and fostering the domestic fishing industry. Meanwhile, delegates to the commission, men appointed by their state governments, flew to Europe and made handshake agreements over tables littered with cigarettes and whiskeys. "The psychology of the 200-mile limit and the nationalism of developing countries is not designed for the management of a world resource like the tuna," John Mulligan, director of an industry-linked tuna association, told a journalist writing in *The Atlantic* in 1976. "It's a very complex problem, one which has been talked about at the Law of the Sea conferences and will be talked about again, but with so many countries involved, it's an almost impossible task."

ON A DARK January morning in 1975, Daryl Schmid rose, alone in her bed, before the sun did. The predawn darkness was so cold that frost had started branching along the framed glass of her two-story split-level, a home she rented from her parents. She peeled off her sleeping clothes, pulled on a pair of slacks, yawned, and started helping her two daughters get ready for school. Once mothering duties were discharged—lunches packed, hair brushed, 12-year-old Janet and 11-year-old Susanne off to school— Daryl slid into her car and headed for the ocean. Her smooth curls bobbing with every pothole, she drove along the road to the north end of the Point Judith Pond, pulling into her regular parking spot at the Ram Point Marina. She key-jingled her way to the front door and opened up her family's business for yet another day.

Once tourists and other gawkers leave for the season, winter can be strangely busy in a maritime town. Gear needs to be fixed and replaced, rotting boards and cracked plastic demand repairs, and the fish don't stop swimming. As long as the harbor remained free of ice and the fishing stayed good, crewed boats continued to ply their trade, and all that meant work to be done at and around the boatyard. So, on that cold January day in 1975, sitting behind the desk in a room she had known for most of her life, Daryl watched the front door to the marina's office swing open as Al Anderson strode into the room.

Closing the door against the midwinter wind, the broad-shouldered man, neatly dressed in a shirt and tie, took stock of the room and walked up to the counter where Daryl sat. He wanted to know who owned a license plate he had seen in the parking lot: Rhode Island 8487. "I do," Daryl said, looking him straight in the eye. "You want to sell it?" he asked her. Those numbers, he said, were the last four of his home phone number, the one he used for his fishing charter business, and he wanted the plate. "No," said Daryl, told him it wasn't for sale. She had inherited the simple plate from her grandmother—its four digits identified it as a classic plate, distinctively old-stock New England—and she couldn't imagine selling it. Al nodded grudgingly and hung around the office for a few minutes before walking down to the boatyard, where he struck up a conversation with mechanic Stu Knowles. Now he had bigger things on his mind than the license plate: he wanted to know more about that gal in the office. "Forget it,"

Knowles said. "She just went through a divorce and isn't dating anyone."

Daryl's father, Bill Schmid, had lucked into opening up the Ram Point Marina: after graduating high school, he apprenticed as a boatbuilder, and in 1942 he joined the US Coast Guard and married his high school sweetheart, Marilyn. At just 19, he had already spent most of his life sailing, his skills on the water earning him a quick promotion to chief petty officer and command of his own boat and crew. After the war, the Schmids moved to a small seasonal community in Narragansett called Breakwater Village with their two small daughters, Daryl and Liane. Here, Schmid fished to support his family in a summer village where they were the only year-round residents. For four years Schmid fished on other people's boats, hauling up nets and arriving home stinking of fish guts, and then found work at a boatyard. There, he struck up a relationship with a customer named Robert Parsons, a wealthy man who loved his three sailboats but didn't love the work of sailing them himself. Periodically, he hired Schmid to take him out and, once the two were close, asked what it would take to monopolize Schmid's time during the summer season. To this, Schmid pointed out the obvious: he needed a way to make money during the winter. So Parsons agreed to finance a boatbuilding business at the tip of the Point Judith Pond, a business that grew into the Schmid family's boatyard and marina. From there, he sold gas, supplies, and parts, all with a handsome view of the bay.

Daryl worked for her parents as their very first "dock boy"

during junior high, pumping gas and watching the store during her lunch break so they could go home and eat. She helped out on weekends, and watched her mother tally numbers and order parts, her father drip sweat and wring oily towels over sputtering engines. She stopped working at Ram Point after she got married and had her two girls, but while her marriage was failing she needed an exit ramp. Her parents bought a house off the new highway and rented it to her; to support herself and the girls, she went back to working at the marina. Her life had been spent a stone's throw from the ocean, and she was happy there.

A few months after his first encounter with Daryl, Al stopped into the marina again, this time looking for a rare long bolt to fasten a used giant tuna fishing chair to his eight-meter Bonita boat. Striding up to the counter, he asked Daryl if she knew what the piece was and if she could find one. Without rummaging around or making a fuss, she calmly pulled out the exact bolt Al needed from the store's jumble of items. This woman, Al thought to himself, was his destiny. A few days later he returned to the marina, awkwardly perusing items and loitering until the office had cleared. He asked Daryl if she wanted to go out. "What do you have in mind?" she replied pleasantly. Al stared at her, stymied. He had prepared for yes and he had prepared for no but he wasn't prepared for a follow-up question. He stammered for a bit before proposing dinner and a movie. "Okay, sure," responded Daryl with a smile. (It turned out that Stu the Mechanic, his loyalties firm, had ratted Al out, so she had anticipated the invitation.)

On their first date, the pair went out to dinner—where they discovered a mutual love of nonfiction biology books—and then went to see Burt Reynolds as a football-playing ex-con in *The*

Longest Yard. They both knew, from that night, that for each other, this was *it*. Daryl often teased Al, accustomed to getting his own way, that if he wanted her license plate badly enough, he had to marry her to get it. They wed 18 months later, her girls in dresses, Al in his best suit; Al said he felt like the luckiest man in the state.

Daryl loved Al deeply, but she was a perceptive woman and already had a sense of the troubling childhood memories Al often struggled under, a weight he rarely shared. Within weeks of moving in together, she had discovered that he could be disturbingly mercurial, ranging from jolly to brooding in an instant. He guarded his possessions carefully and snapped at the young girls if he thought they were handling them roughly. Once, he took his stepdaughters to the park so they could watch him fly a balsa wood glider he'd built. Soon after arriving, the glider ended up slightly damaged, and Al, brooding and snappy, immediately ended the outing. Susanne, his younger stepdaughter, noticed that, particularly when embarrassed, Al would overreact at the smallest perceived slight. Once, though, he pretended to be a sea turtle, flapping his arms as if they were flippers underwater, and the entire family broke out laughing. Susanne learned that if Al seemed on edge, she could ask him to reprise his sea turtle impression; he'd usually soften even after his most half-hearted attempt at transforming into a turtle, arms flailing at his sides.

Al's troubled childhood also manifested in more obvious ways. During Al's first Christmas with Daryl and the girls, the four went to visit his wife's parents as a blended family. That first morning, everyone in Daryl's family's had left Al presents under the tree, but he had no idea how to react to them. Sitting in the living room that Christmas morning, he handled a box with his name on it for longer than normal as her family looked on. To break the awkward silence, Daryl's sister finally said, "Alan, you've gotta open your present." After a pause he said, "It's mine?" His daughters rolled their eyes as he slowly, carefully, shook the box, turned it over, wondered what was inside, marveled at it, carefully unwrapped it, shook it some more, then finally opened his present. After his first disastrous foray into marriage and so many years living alone with his rods and books, Al had simply settled into a life without presents. He later confided in Daryl that his parents had never given him a single true Christmas present, not the way that other kids got them.

Once, Daryl asked Al about the toys he had when he was little. He thought for a long time, until a distant memory emerged: there was one cigar box decorated with a stripe, and he used to drag it behind him on a string. This was his only memory of owning a toy. Al's father had once bought a toy train for Christmas but Al wasn't allowed to play with it: every year Arthur set it up around the Christmas tree, where it looped its endless, predictable trajectory, never touched by the small boy who had unwrapped it.

At parties, despite being tall and handsome, Al hesitated to socialize, tending to hang back in the corner until someone approached. He could be hard to get talking, although his stepdaughters soon learned that engaging him on topics he could hold his own on, such as fishing, weather, and biology, was a guaranteed way to break the ice. In small ways, the damage lurking within Al's structured days of teaching and fishing erupted, often out of the blue. His teenage stepdaughters learned to steer clear of Al's office, of his fishing equipment, of touching anything that could prompt a harsh, inexplicable rebuke. This quickness to anger closely resembled the fury Al could unleash at his own tuna-fishing crewmembers and clients, and eventually became part of his professional reputation. If you went fishing with Captain Al, you were guaranteed to catch fish, but you were equally likely to get yelled at if you messed up.

Still, on other days, his enthusiasm for fishing, science, and ideas washed across his family in an effervescent wave. Once, Al decided to dress up for one of his classes, and the girls helped him cut and decorate a box that he pulled over his tall frame. He arrived at school that day as "Mr. Cell," wearing a black polo shirt buttoned to the neck, his dark-framed glasses and black sideburns setting off the black-square-painted cell that contained all Al's microscopic innards. In a photo from that day, his arms poke out from holes cut in either side of the box, elbows pointing outward and hands pointing toward the dashed lines of his carefully drawn mitochondria, chromatin, and ribosomes, his mouth wide in a silent, gap-toothed cheer. Living with Al was often an irrepressible joy.

Once school resumed in the fall, Al headed out on overnight striper fishing trips throughout the season. After a long night on the North Rip, he'd slouch into his classroom, always on time but often exhausted and irritable. Just as his home life with Daryl and the girls ranged from ecstatic to chilly in an instant, Al found himself both loved and despised by his students. His enthusiasm for preparing them for careers in science often veered into the extreme, and he didn't give easy grades. He designed his human biology curriculum for seniors on what he wished he himself had learned before heading to university, and he demanded that his students buckle down in the class. The seniors, many of whom wanted to pursue careers in medicine and biology, dissected cats in the class, and Al pushed them, deducting marks when they, for instance, labeled a leg bone only a tibia, when it was, as Al chided, the left tibia. "There happens to be two tibias," he lectured as the teens rolled their eyes behind his back. Yet for every student who, as he said, "hated his guts," Al found the parade of critiques paled with the validation of chatting with the small handful of graduated students who returned, gushingly grateful at the extent to which they had entered college so well prepared.

Out on the water, during weekends, evenings, and the summer charters he ran when school was out, Al preferred running his boat solo: racing from captain's chair to the rods, spinning tales of fish behavior and past exploits for clients, furiously yelling at other captains if they dared draw their boats over his lines. Sometimes, if chartering a big group, he'd hire a mate, including, for one season, his stepdaughter Susanne, to help with the rods and landing fish, and his ire wasn't reserved for fellow skippers. One former mate, Steve Tombs, who fished with Al as a teenager, recalled Al snarling over the shoulders of startled, wide-eyed clients. "What are you doing? Didn't your mother teach you anything?" he'd growl. One morning Al accused another mate, Matty Di-Matteo, of forgetting to pack the right rods, and the two got into a screaming match in front of a boatload of terrified onlookers. DiMatteo smashed Al's prized flashlight against the side of the boat while Al bellowed. A fishing buddy once asked Al if it bothered him that so many people seemed to hate him. "William," Al growled in response, "I love it." Even many of Al's friends started as enemies. Captain Ed Everich first met Al in the 1960s when the pair tangled fishing lines. They screamed obscenities at each other, but after Everich bought Al some new gear, the breach was mended. "You either liked him or you didn't, and I think the feeling was mutual," Everich said.

Yet whatever Al was doing had been paying off. He closely observed gear and techniques other captains used to land fish and even adapted a new "umbrella frame" rig of wire and latex tubing, used to simulate an eel-like forage fish called a sand lance, to his own specifications. If he twisted the wires in a corkscrew, they bobbed and swayed with the currents, closely mimicking the fish's undulating swimming. Drawn by guaranteed catches and word of mouth, more charter clients were calling and booking repeat fishing trips year after year. For every fisherman who avoided Al, there seemed to be two or more who sought him out, eager for his blustering camaraderie, work ethic, and knowledge of local waters. Some were also curious about the red-and-white T-emblazoned triangular flags, each one representing a fish he had tagged and released that day, that he raised up on the *Prowler*'s outriggers out on the water.

Al, who had tagged hundreds of striped bass since his time in graduate school, heard about his first tagging return, or "recapture," in 1970. He received the news via a typewritten letter that included a patch—a gold-stitched fish on a royal blue background denoting his first recaptured tag for the American Littoral Society—as well as the original tagging card he had filled out and mailed after initially tagging the fish, the mustard-yellow spaghetti tag he had embedded in its back, and a handwritten note from the angler who had recaught the fish on the shores off Montauk two months after Al had caught and tagged it off Point

Judith. Chuffed, Al carefully placed the entire package in a plastic sheath as proof that his tagging work meant something, and was being reflected back by the wider world as having value. As he encouraged his clients to tag fish, every year a few more fishermen were willing to let fish go in the name of science. It was here that a larger picture began to reveal itself: he had discovered a niche in the market for other curious fishermen who wanted to tag fish and see where they ended up.

Al's favorite fishing spot, a rushing of currents known as the North Rip that lies off Block Island's northern tip, had been formed by the southern edge of the region's last retreating glacier. Nineteen thousand years ago, the glacier ground up and deposited clay, sand, and gravel in a roughly triangular shape along its southernmost border, or terminal moraine. As it melted, it left barn-sized chunks of ice atop Block Island, which was then about six times as large as when Al first encountered it.

During the last ice age, the port of Point Judith lay underneath a massive sheet of glacial ice that perched atop the absolute edge of the North American continent. As this icy topper melted and scraped southward across the sandstone and shale bedrock, it sheared steep mountains to rolling hummocks while deepening and smoothing its valleys and lowlands. Everything from Narragansett downward was peninsular, divided from the mainland by the long, jagged Point Judith Pond. Viewed from above, the scraped terrain recalls honey dripping from a bottle's wide mouth, a carved geologic north-to-south recollection of migrating glaciers. A few thousand years after the ice melted, a surge of molten lava bubbled up, creating a giant seam of pink crystallized rock known to stonecutters as Narragansett Pier granite.

Indigenous families lived on the land now known as Rhode Island for thousands of years, the eventual devastation of their people leaving little upon this land but artifacts: quartz and green shale arrowheads, hide scrapers and drills, stone axes, soapstone bowls. Once, they lived here year-round, moving to winter settlements when storms and winds threw snow pellets, lashed the rocky shores, and bent the clinging pines nearly sideways.

Al also dug deep into the history of settlement along the coast. By the 15th century, Narragansett tribes lived in a handful of semipermanent villages, governed by chiefs called sachems. They had cleared land on which they grew corn, squash, beans, and pumpkin in brimming summer gardens. They harvested hickory and chestnuts—the latter would be wiped out by disease—pounding flours and pressing oil that overwintered in clay pots. From the ocean and its rivers, they harvested spring runs of alewives and collected the coastline's abundant, seemingly inexhaustible shellfish. They made ceramics and were remarkable stonemasons. They were overwhelmed by disease, cheated of their land, and largely massacred in skirmishes and battles with British colonists by the 1700s. "The natives received the voyagers kindly, bartering furs, provisions etc.," reads one 1902 account. "But the [colonizers] nonetheless endeavored to steal women and children to carry to Europe, and thus early began that bitter enmity." By the time those words were penned, white settlers had claimed their land, parceled it out, and spread livestock across the rocky region.

America's Revolutionary War with the British came to the area in 1775, when batteries were built along Narragansett Bay to prevent the passage of unwanted ships. By then, the region had become famous for livestock, including Narragansett Pacer horses, which had the smoothest gaits over New England's rough roads, and dairy cattle—one farm's 100 cows produced more than 6 tonnes of cheese annually. As builders erected piers and wharves along the coast, shipbuilding and fishing for alewives, bass, smelt, and white perch became mainstays. The region was notorious for unpredictable weather: on July 14, 1894, an eight-minute hailstorm destroyed $50,000 worth of property (nearly $2 million in today's dollars) and caused "many cases of nervous prostration." In summer, fashionable city crowds flocked to Narragansett's beaches, but in 1900 a giant fire consumed the Narragansett Pier Casino, the town's primary landmark, along with a large hotel and several blocks of buildings. By 1920, the town's population hit a low of 1,000.

As new highways branched like roots across New England in the postwar boom, the town reached 3,500 people in 1960, steadily drawing aspirational families, like Daryl's, to its safe harbors. For decades at that point, the Atlantic Tuna Club[3] had been attracting anglers who wanted to fish the hard-fighting small-school bluefin every summer. Learning this history and lay of the land was one method Al used to make a name for himself charter fishing, regaling clients with stories of the local fishing and geology so they came back for more. He headed out as frequently as he could, fishing with self-assured swagger alongside grizzled veterans and full-time skippers, trying to understand, as fast as he could, where the fish were and how to catch them.

In the mid-1970s, as striped bass catches off New England

boomed, Daryl quickly developed a reputation as the mysterious woman who dropped huge stripers off at the fishermen's co-op, the group her father had cofounded decades before, for sale to the public. As dawn broke, Al would arrive home from his early-morning fishing trips, his big Chevy half-ton truck loaded with fish. He'd squeeze in a quick shower and breakfast, throw on his slacks, white shirt, and tie, and they'd head to Westerly High in Daryl's car. She, on the other hand, managed getting the fish sold.

Intrigued by Al's growing reputation, an editor from *The Fisherman* magazine, a respected regional fishing publication, contacted Al to see if he'd be willing to contribute fishing reports about Block Island; Al jumped at the chance to get his name out there. If he could make some money *and* spread the gospel of fish tagging—of stripers, bluefin, billfish, or sharks—all the better.

Despite ongoing concern in the scientific community that over-fishing was hurting Atlantic fish stocks, fishermen continued to pull record-breaking hauls of fish out of the ocean. Those fish provided anglers with anecdotal fodder that everything was fine. In 1979, while fishing out of Port Hood, Nova Scotia, a Canadian angler named Ken Fraser caught the largest bluefin tuna ever recorded. In a photo taken after the catch, the 1,694-pound fish dwarfs Fraser's heavy, water-splattered frame; the width of the bluefin's sickled tailfin appears longer than Fraser's entire torso. He holds a huge-reeled tuna rod in his right hand as his left rests on the fish's sharp dorsal fin, his eyebrows slightly raised over wide eyes. Beside him, the fish, which was likely around 25 years old, gapes in death, hanging by a rope from an off-camera winch. The photo represented a strange landmark: demand for tuna meant that no fish of its size would ever grow that large within

Fraser's lifetime and perhaps ever again. Since 1973, American sport anglers had been allowed to catch only four bluefin a day, only one of which could be less than 14 pounds or heavier than 115 pounds. Yet as commercial fishing continued apace on both sides of the Atlantic, even respected longtime tuna fishermen like Al Ristori were increasingly convinced the species was, as he put it, "seriously overfished."

This was a phenomenon that also extended to Al Anderson's beloved striped bass. In 1979, recreational fishers landed more than a million total of the fish, and Congress directed the US Fish and Wildlife Service and the NMFS to commission a multimillion-dollar emergency scientific study to investigate the dire state of the bass fishery between North Carolina and Maine. Despite Chapman's apparent trust in the basic premise of maximum-sustainable-yield-based fisheries management—that regulators could respond to scientific data quickly and effectively—it took five years of dwindling stocks until federal conservation limits were passed.

As charter captains bucked under new catch limits for stripers, Al sensed a growing opportunity. Sure, it wasn't legal to catch and keep small bluefin, but tagging them offered a convenient loophole and provided a chance to study the fish more closely. Sometimes when he caught a particularly handsome large bluefin, he would legally sell it to one of the Japanese buyers on the docks. It seemed crazy, the prices they offered for bluefin, and there seemed to be no ceiling on how high the fish's value might rise. And while Al didn't know it at the time, why and how this occurred—and how a previously worthless fish captured the imagination of a generation's worth of northeastern fishermen—drew its foundation from events decades prior. They would culminate in the arrival of one of the world's most notorious cult leaders and a loyal following of his acolytes in one of America's most historic fishing communities, while setting the stage for the tempest on the horizon.

CHAPTER FIVE

RISING MOON, FLYING FISH

~~~~~~~~~~~~~~~~~~~~~~~~

Japan, the 1970s

It is better to go home and make your net than to gaze
longingly at the fish in the deep pool.
—JAPANESE PROVERB

I n 1971, Japan Airlines Co., or JAL, faced a conundrum. For
every six tonnes of freight transported halfway around the
world to North America by its giant cargo planes, it was only
shipping less than a tonne back. So the company tasked a 41-year-
old Japanese employee, Akira Okazaki, with filling the empty
space. That meant Okazaki needed to find a high-value perishable
product that justified the high cost of flying it to Japan. Surging
demand for bluefin in the Japanese restaurant market seemed like
a good opportunity, and Okazaki had heard rumors that the biggest, fattest fish could be found off Canada's east coast—reputational
remnants of Wedgeport, Nova Scotia's formerly famous fishery.
Okazaki transmitted a teletype to the airline's first Canadian

employee, Wayne MacAlpine, asking him to investigate the opportunity. MacAlpine did, and eventually got a tip to check out North Lake, Prince Edward Island (PEI), a tiny village perched on a harbor of the country's smallest province, then known mostly for its tasty potatoes and the storybook character Anne of Green Gables.

In North Lake, MacAlpine found a town equally blessed and plagued by giant bluefin. During the summer, sportfishermen arrived from around the world to catch the warm-blooded monsters, which often tipped the scales at more than 600 pounds, on rod and reel. Yet some cod-catching islanders hated the fish, which shredded nets and ate fish they were trying to catch, and whose streaks of rich belly fat rendered the fish, to their tastes, completely inedible when cooked. They were worthless at the time, MacAlpine discovered, so, as was done in nearby Nova Scotia, sportfishing clubs had to pay someone to remove the fish, dig a hole, and bury it before it rotted on the dock.

MacAlpine and Okazaki knew the town had the potential for pay dirt. But first they had to surmount the problem of how to transport the fish from a tiny village to the nearest airport that could accommodate JAL's giant jets: New York City's JFK, 1,300 kilometers away. It was obvious the fish had to be trucked, but how could they possibly move the delicate, perishable fish without the bruises, scrapes, and rotting that would render it worthless in Tokyo? That called for the highly specific skills of a local tough-guy businessman named Albert Griffin.

Every time MacAlpine arrived on the island from his office in Toronto, Griffin, a former bootlegger and rumrunner, picked him up at the airport, threw him the keys, plopped down in the passenger seat, and pulled out a little brown bag of hooch. Griffin was a tough, exacting businessman, and MacAlpine initially worried he might clash with the JAL officials, with their pinstripe suits and impeccable manners. But over long nights of beer and cribbage (during which Griffin made it uncomfortably clear he couldn't tell the Japanese men apart), the men landed on a common language of money.

Using a profit-sharing system proposed by the airline, local fishermen would receive a cut for the auction price of each fish they caught and shipped, which encouraged them to catch, cool, prep, and deliver the highest-quality fish possible. To ship the huge fish, Griffin approached the Dingwell brothers, two funeral directors who also built coffins. Those coffins, when filled with chipped ice, made for a cold, bump-proof system to JFK that would satisfy even the pickiest auction buyers.

Okazaki ran some early experiments on cheaper yellowfin tuna, filling their inner cavities with piles of ice, wrapping each body in plastic, putting that into a vinyl bag, and then spraying the whole package with a thin layer of the insulating foam urethane to contain the chill. The first fish JAL shipped using this method sat for three days in ice, cost $1,000 to truck to New York (about $7,000 today), traveled for 14 hours in the hold of a DC-8 airplane, and sold for a disappointing 80 cents per pound, mostly due to poor refrigeration en route. Okazaki knew they could do better and convinced his bosses they needed to custom-build a refrigerated container that would fit in the airplane's lower cargo hold.

On the dawn of August 14, 1972, five large Atlantic bluefin arrived at Tokyo's Tsukiji (SKEE-gee) fish market four days after they had been pulled from the chilly waters off PEI. The first fish sold for $18 a pound that day—about $130 a pound, or $52,000 for a single tuna, in today's dollars. The day became known at the Tokyo market as "the day of the flying fish." Before the end of that year's tuna season, JAL had shipped and sold nearly 200 fish at auction, and by 1976, Okazaki and his colleagues were importing around 1,000 fish every year. Not to be outdone, other cargo companies and tuna middlemen, hungry for profits promised by flipping the basement-price western Atlantic bluefin, joined the fray, and a new industry was born.

For fishermen across the northeastern Atlantic coast, it seemed like Japanese tuna buyers and "tuna techs" started arriving on docks virtually overnight, hoping to land the best fish as early in the fishing season as possible and at the lowest prices. They showed up gradually at first, then all at once, and they often arrived accompanied by translators. They waited patiently on the docks, eyes pinned to the horizon in their pressed, high-waisted khakis and crisp white shirts.

After catching a 300-pound fish and selling it for the 25 cents per pound offered by many Japanese buyers in those early days, earning $40 per fish (more than $200 today) felt like a good deal, former tuna fisherman Charlie Donilon told me. And demand pushed prices higher quickly, within months: first from 25 cents a pound to 30 cents, then eventually climbing to as high as 75 cents per pound. The day the first fisherman got that price for his fish, a notorious Rhode Island tuna fisherman named Don Slater got on the VHF (very high frequency) marine radio and announced into the New England breezes: "You mark my words," he crowed. "We're going to see a dollar a pound someday."

While demand for US- and Canadian-caught bluefin kept those

prices climbing, within the decade, a sushi craze had begun ricocheting across the world. Diners from Peru to Germany, Ohio to Tallahassee were ordering *maguro*, or bluefin, by the platterful. Sushi was not only changing the world; it was also emblematic of a changing world.

THE ADVENT OF sushi in Japanese culture, like that of wine, first started as a preservation method a millennium ago. To guard against hunger, early Southeast Asian farmers gutted and packed fish, often whole, in sealed jars of cooked rice, where the flesh fermented into a funky, lactic mass. The fish remained edible for up to a year, although the farmers threw the rotted, desiccated encasing rice away. By AD 800, this *narezushi* technique had arrived in Japan (*nare* means "become"). By opening those fermenting jars early, a handful of adventurous diners figured out that they could eat not just the fish but also the tangy rice that encased it. With the popularization of rice vinegar in the 1700s, some chefs realized they could make vinegar from sake lees, then mix it with rice and serve it alongside raw fish for a result similar to the jar-fermented style.

According to lore, a chef named Hanaya Yohei first popularized the modern-style preparation of sushi in Tokyo in the early 1800s, selling pressed balls of vinegared rice and *neta*, or prepared fish, out of a box he carried on his back. At that time, most Japanese diners avoided eating tuna and held particular disdain for *toro*, the fatty belly meat of the red-blooded maguro. In the 1840s, fisher-men caught an unintended glut of the fish and offloaded them cheaply to street vendors serving Tokyo's lowest classes. These vendors soaked bluefin steaks in soy sauce to mask their metallic tang in an attempt to make the *gezakana*, or "inferior fish," more palatable. And for those who couldn't stomach the taste, there was always, like the earliest sushi, the option of fermentation. They called the bluefin preparation *shibi*, or "four days," for the time it took a bluefin buried in the ground to ferment to the point where its rich, ferric-tasting flesh became palatable. (One other inter-pretation of the meaning of *shibi* is "the day of death.")

Another lighter-fleshed species of tuna, yellowfin, began to catch on in Japan in the late 1800s. An early woodblock print by artist Utagawa Hiroshige III, in his 1877 series "Famous Prod-ucts of Japan," depicts a fall scene of five shallow-keeled boats manned by grass-skirted fishermen in white headbands, spearing yellowfin out of the sea. In the center, around a writhing swarm of netted fish, one white-bellied tuna jumps into the air, fins out, back arched. A description of the scene, in classic Japanese script scrawled in the print's top-left corner, describes fishermen draw-ing nets attached to two fishing boats upward as an observer watches from a hilltop tower. When the net was full of tuna, the man high above the water would wave a flag, and the fishermen would draw their spears and throw, flooding the seawater with blood.

In the years following the American occupation of Japan after the Second World War, Japanese diners began to seek out the fatty red beef that the strange foreigners had enjoyed so much. That trend led to an increasing acceptance of the fatty red-fleshed bluefin served as nigiri or sashimi in the country's hundreds of Japanese restaurants. In April 1958, a Japanese inventor named Yoshiaki Shiraishi founded the first conveyor-belt sushi restaurant in Osaka. He had found inspiration watching bottles of beer smoothly navigate a brewery processing line and worked with a 20-person machine shop crew to bring his *kaiten-zushi*, or "turnover sushi," dream to life. When the business launched with prices nearly one-third cheaper than those of his competitors, lines of customers snaked out the door, and the concept expanded across the Japanese islands.

The growth in sushi's availability and affordability, paired with the introduction of modern refrigeration and low-temperature freezing that preserved bluefin's quality, meant Japan's dining public quickly developed a ravenous appetite for bluefin that would *torokeru*, or "melt away," on the tongue. Japan's airline industry ran marketing campaigns to support the spread of their premium bluefin products, according to *The Story of Sushi* author Trevor Corson, and soon the craze for maguro had crept across the country. Sitting on diner stools and on wooden benches, Japanese diners dipped the unctuous raw bluefin meat in sweet soy sauce, pairing it with paper-thin slices of sharp pickled ginger and the white-hot heat of wasabi. Then they ordered more.

In August 1971, a purse seining captain named Frank Cyganowski[1] caught the global seafood industry's attention when he bought bluefin from local fishermen and transferred those fish, at sea, to the Japanese freezer vessel *Kuroshio Maru 37*. The giant ship had been built to carry 272 tonnes of cargo and could chill about 9 tonnes of fish to minus 45 degrees Celsius in a day. As part of the transaction, Cyganowski also bought bluefin from local fishermen for 10 cents per pound—not revealing that he was selling those fish to his Japanese buyers for more than twice that, at 48 cents per pound. The ship left Cape Cod Bay in October with a load of frozen bluefin and arrived in Japan a few months later. The next year another Japanese ship arrived, except this time its owner was willing to pay up to $1 per pound. That meant one fish that was once sold for $40 (or about $280 today) by an early bluefin fisherman like Al Anderson or Charlie Donilon was suddenly worth up to $1,000 (or $6,600). For a businessman able to buy top-quality bluefin at a US dock and then fly or ship the whole, headless, gutted fish to Japan for auction at the world's largest fish market, there were fortunes to be made.

ONE MAN WHO played a part in those fortunes was Allan Hokanson, a devout Lutheran Christian who grew up in Seattle. In 1966, the 18-year-old Hokanson spent his summer working in Alaska aboard the *Redwing,* a tublike 18-meter-long galvanized steel tender boat that picked up fish from other boats and brought them back to a cannery. The work was hard. He was a runner and in good shape, so the hard work felt manageable, and he earned enough money that summer to enroll in college, where he studied oceanography at the University of Washington. That first year out on the ocean, Hokanson wrote a letter home recounting falling in love with the sea: navigating icebergs, watching whales spout and porpoises frolic, and finding himself "surrounded by jagged peaks, which rose in all their greatness and splendor above the shining sea." In 1972, at the end of the fishing season after graduation, the 24-year-old idealist took his fishing cash and decided to travel to Europe. He stuffed a large red pack with clothes, two cameras, camping gear, and his most prized possession: a new Bible he bought with his birthday money.

Starting in England, where he slept under the stars some nights and in youth hostels on others, Hokanson stretched his budget by eating cheap local cheeses and hunks of bread while drinking strong, creamy English tea. He stopped in Wales and Scotland, and eventually took a hovercraft across the English Channel. In France, he couldn't speak or understand the language, which he hated. Lonely and isolated, he hopped on a train and headed north.

Forty days after leaving Seattle, Hokanson disembarked from another train in Lübeck, Germany. As he walked away from the station, a young woman—speaking perfect, British-inflected English—approached him with an offer. Would he like to come to a meeting about religion and science? "Why not?" Hokanson thought; he had nothing else to do. Alongside two young German women, he spent two days reading through an English translation of the Reverend Sun Myung Moon's teachings. At night, he slept at a youth hostel, waiting for a sign from his God that Moon was truly the messiah returned. He spent a third morning sitting in a local Christian church, comparing passages from his Bible with those in the Unification Church's text. The two volumes played off each other, he thought, which he took as a sign. They both saw so much beauty in the natural world and in God's word. Perhaps this was his destiny calling.

A few hours later, after finding out that the mobile missionary team was leaving town, he made his choice. Clutching his cameras but none of his belongings, which church members promised they would gather and send, Hokanson climbed into a van alongside all the other fresh-faced, fired-up teenagers and 20-somethings. Behind the van rocked a trailer that tripled as an office, sleeping quarters, and prayer meeting room. Hokanson was officially the newest member of the Unification Church of Germany.

In 1950, in the uncertain days following the Second World War, a South Korean preacher named Sun Myung Moon and a friend of his built the Unification Church's first informal chapel in Busan, South Korea, out of discarded US military ration crates. The religion was formally established in Seoul in 1954. With the self-anointed Moon at the helm, the group rapidly expanded across South Korea and Japan and eventually Europe, funded by millions of dollars' worth of profits from herbal remedies, candles, nuts, tiny religious statues, and flowers sold by acolytes who stood on street corners and went door-to-door. At the church's core was the Divine Principle, a series of teachings based on notes Moon had scribbled in the margins of his Bible—as well as on the walls and ceilings of his home—predicated on the belief that Moon was born to complete Jesus's work on earth.

Moon first arrived in the United States in the predawn hours of February 1965 bearing a briefcase of Korean soil. He had flown overnight and arrived at San Francisco International Airport with the goal of consecrating dozens of sacred sites across America. Over the next six weeks, he and a group of disciples, some of whom had lived in the United States as early as 1959, crisscrossed the country in a blue Plymouth Fury station wagon, burying Korean soil and stones and blessing 55 individual locations, including a hill overlooking San Francisco and the grassy lawn in front of the US Capitol. In his theology, America was the "archangel nation" that would help free his homeland from the scourge of the Communists, who had beaten and jailed the young preacher for nearly three years following accusations of polygamy.

In 1971, Moon permanently moved to the United States, and his treasurer's assistant arrived soon after with 1.8 million undeclared dollars, reportedly donated by Japanese acolytes, stuffed into another briefcase. It was around this time that a German Unification Church leader offered Hokanson the chance to travel back to his homeland as part of a 30-person missionary group. The leader, looking sternly at Hokanson, asked him if he was serious about staying with the church and serving, or if he planned to head home if the church paid for his ticket. Hokanson decided to stop fighting destiny and arrived at the church's Tarrytown, New York, headquarters that February as a full convert.

Within the year, Hokanson had been selected for leadership training, a rare opportunity to personally meet and spend time with Moon. While the young fisherman took part in a group celebration, Moon summoned his acolytes to gather around under a big tree. He asked everyone who had experience on the water to stand up, and about two dozen, including Hokanson, did. Some could row boats, some had worked in the navy, but no one had the type of experience Hokanson did. Moon ordered him to renew his boating license, and his new life as captain for Moon's brand-new sportfishing boat, the *New Hope*, began.

Their first summer out on the water, a Freeport tackle shop

owner told Moon and Hokanson a story about the 1,000-pound bluefin that schooled off Massachusetts's shores. Moon adored the idea of fighting a giant fish and immediately bought a tuna-fighting chair for the *New Hope*'s deck and six heavy-duty tuna fishing poles, each with a winch-like metal reel.

In July 1974, Hokanson piloted the *New Hope* for Moon and a handful of unpaid working church members for their first trip to Gloucester, Massachusetts. The group rented a room at the Rockaway Hotel, a grand old dame of a beach resort near the town's Rocky Neck, and moored their boat at the hotel's dock. For two weeks the group fished but didn't catch anything. By the end of the first season, the inexperienced crew hadn't landed a single tuna. But Moon was hooked. "When the sun starts to sink at the end of the day, I feel like ordering the sun to stay up so I can stay out and get another strike," said Moon in a speech a few years later. "Though you lose the tuna, you just keep trying until you catch one. That's the way it goes." For Moon, the meaning of fishing lay in the attempt of fishing—the faith that a fish would come—and not necessarily in its guaranteed success, and disciples like Hokanson thrilled at Moon's certainty.

The next year the group came back, rented rooms at the same hotel, and set out once again for tuna. This time, at least, they started hooking them, although they struggled with the task of actually landing one. On the third fish they hooked, Hokanson turned around to find Moon and one of his senior leaders trying to tie the line of a second fishing pole onto a line that was already dragging a huge tuna about 30 meters behind the boat. "What's going on here?" Hokanson asked himself in disbelief as the fish slipped off the sagging line and loosened hook. The group had fished for 21 days and hooked 16 fish but didn't land a single one. Then the next day, to their joy and elation, they caught two, which the entire crew hauled aboard with hooks and grunts. Before motoring toward the town's fishing port to unload the giant fish, the group stopped triumphantly at the Rockaway Hotel's dock so Moon could show his wife and children the fish. Someone even ran into the hotel's restaurant yelling that the *New Hope* had landed a tuna. Nearly all its diners came out to see the fish; so many people climbed onto the dock that it started to sink. Reenacting this scene in my mind, I can hear the screams and yells of children, the jostling to get closer to the fish and to touch the bluefin's metallic skin, their fingertips on its fins' sharp points.

Everything about bluefin exhilarated Moon, from how the fish tucked its fins close to its body like a state-of-the-art Concorde jet to the fish's shimmering beauty and intelligence. Like Al Anderson, he was entranced by the fish's life cycle and ability to prowl the world. "If the ocean were divided into sovereignties, then the tuna fish would be in trouble because it would need a visa!" he once exclaimed. Moon believed that all sport-minded men like himself—those men of an earlier age like Zane Grey and Ernest Hemingway—coveted the chance to catch a big tuna, that the catching of the fish was a symbol of charisma.

Over the next eight years, the *New Hope* annually caught about 35 giant bluefin every season, often catching their one-giant-tuna daily quota set by the federal government before noon. "No other fishermen could come close," said Hokanson of their exploits. The biggest fish they caught was over 1,000 pounds. Another day, an

880-pound fish pulled the *New Hope* around for five hours, covering nearly 16 kilometers. For Hokanson it felt like having a semitrailer on the end of his fishing line.

Moon, who had started calling himself the "King of the Ocean," then put the next phase of his world-spanning plan into motion. "I have the entire system worked out, starting with boatbuilding," he said. "After we build the boats, we catch the fish and process them for the market, and then have a distribution network. This is not just on the drawing board; I have already done it." Many of his followers had never fished before, so Moon decided to try a new training method: instead of relying on a big, hard-to-maneuver boat like the *New Hope*, he thought a smaller vessel with a simple outboard engine might do better. And that was just the beginning. Like any respectable historical leader, he wanted a fleet.

Moon's dozens of new 8.5-meter-long white fiberglass boats, all named *One Hope*, were manufactured by Master Marine, a Unification Church–associated boatbuilder in Long Island City, Queens. The smaller crafts, which were later manufactured with inboard diesel engines, could be pulled around by a tuna for hours, wearing it out until the fishermen hauled the fish up to the surface with a simple hand line: essentially a long skein of rope coiled in a basket. Tied together at the docks, these bobbing crafts, with their eerily identical names, resembled beehive cells. On celebration days and formal occasions, Moon required Hokanson to don his smart, dark, navy-style uniform, with two rows of vertical gold buttons and a white-and-black captain's hat with crossed golden anchors on its front; he also insisted Hokanson festoon his cap with the five stars of a navy admiral. A picture from this time, yellowed with age, shows the sandy-haired Hokanson, dressed head to toe in rumpled khaki, proudly beside a giant bluefin, with Moon on the fish's other side.

Despite fishing alongside each other for years, the relationship between church fishermen and established skippers from across the northeast was awkward, and often upsetting to well-intentioned acolytes like Hokanson. When Moon and his caravan of followers had first arrived in Gloucester, many of its suspicious locals didn't know what to make of the church members. Some jeeringly called them "Moonies," and some established fishing families in town refused to do business with them. Some would also pull down their pants and show Moonies their butts—"moon" them, as it were—while yelling, "Moonies suck!" One church member originally from the Bronx told the others to stick up for themselves. "When they yell 'Moonies suck!' you yell, 'Your mother sucks! Your sister sucks!'" he told them sagely. Once, a church member did exactly that, which initiated a shouting match over the water: "Your mother sucks!" "No, *you* suck!"

In another, more serious instance, Hokanson was tuna fishing out on the *New Hope* when news crackled across the VHF marine radio: the independent fishing vessel the *White Dove*, its captain Mike Genovese announced, was taking on water. The *White Dove* and *New Hope* were identical, the only two of their kind fishing for bluefin out of Gloucester's busy commercial harbor. Town gossips considered the look-alike boats archenemies: one day they'd tangle lines, another they'd head for the exact same small patch of ocean. Hokanson had an extra gas-powered pump that he kept in his boat for emergencies. Picking up his radio mic, its dark, spiraling cord dipping in time with the rocking waves, he offered to drop off their extra pump. But the *White Dove's* captain said he didn't need Hokanson's help, and eventually Genovese piloted his boat back to the harbor. Hokanson felt relieved, as it would have been a pain to pass off the pump, but still, I can't help wondering: Would Hokanson's help have been accepted if he weren't a "Moonie"?

By 1980, the Unification Church, using corporations affiliated with Unification Church International Inc., had bought $2 million worth of property in Gloucester, including 3 acres of waterfront property and a 20-room Tudor-style mansion nestled on 11 acres. Previously a Catholic-run nursing home, the 1905 villa even had a high stone turret on the property. When the mansion first went up for sale, Gloucester Catholics had prayed, at public masses, that the Moonies wouldn't buy the property. "I'll tell you what I don't like about the Unification Church: they use kids [for free labor] and they're not going to do it in Gloucester," said city mayor Leo Alper. He even sent a personal letter to Pope John Paul II, imploring him to intervene. "I'm no bigot," he told the *Washington Post*, "but I don't want them here." To Alper's rage, the church also arranged to buy Bob's Clam Shack, an erstwhile take-out restaurant, for $650,000 at bankruptcy auction. Moon renamed the eatery the New One, and its acolytes operated it alongside a neighboring fish- and lobster-purchasing business that was the only facility in town equipped to buy local lobster catches. "Nobody comes in and purchases what they're purchasing without a plan," exploded Alper upon hearing news of the sale. "I think the public should know what the plan is. We're a city, and I'm beginning to feel like we're being invaded."

Groups of protesters, holding signs, took to picketing in front of the restaurant and lobster pound. Residents affixed bumper stickers on their cars, protesting the group, including one that said, "Save the human race. Punch a Moonie in the face." One night, they hung an effigy of Moon beside the harbor. Screaming bystanders sometimes threw rocks at members' cars or the boats they were towing. Once, a livid fisherman started throwing chum at one of their boats before ramming it and jarring the Unification Church members aboard.

The hatred some locals harbored for the Moonies hadn't felt dangerous until one evening in 1981 when they attended a community meeting at the Seagull Restaurant about the Unification Church's bluefin fishing fleet and the church's activities in the town. It had been a tense meeting, and they got into the van, relieved to be heading home. A big, calm German named Gerhard Peemoeller, who worked as Moon's personal bodyguard, sat at the wheel. As Peemoeller drove, he realized, with dawning dread, that the van's brakes weren't working. He managed to stop the vehicle, and one member pointed out a pool of fluid under the van: a vandal had cut its brake lines sometime during the meeting. "They tried to kill us that way," Peemoeller wrote in a passage later published in a Unification Church history book. "That is how Gloucester treated us."

WAITING ON THE marinas' gray, salt-whipped planks across the northeast, Japanese buyers wielded small coring devices that would punch a neat, round flesh core from the tuna. Smearing the crimson, vascular meat across their fingers, or laying it on a table, they examined fat content and color. They used meat hooks, graceful arm's-length poles with curved iron points on the end, to move and maneuver the giant fish. Those hooks were unlike any these New Englanders had seen; Americans used two-by-fours with nails hammered through one end. If a buyer liked what he saw, he yelled "Japan fish" and marked the animal's bulging, gun-metal side with a black permanent marker. Local workers wheeled it into a packing room, dumped it into a box filled with ice, nailed it shut, and loaded the box on a refrigerated truck. With this system in place, fish caught near ports could land on the floor of Tokyo's Tsukiji fish market's tuna auction within days.

There were challenges, of course. For one, some of the fisher-men relished the long, brutal fights with the fish out on the open water, but any decent buyer knew these fights burnt out the meat. Unlike other fish species, which lose heat as cold ocean water passes across their gills, tuna have a built-in heat-recirculating system that allows them to regulate their body temperatures. This rete mirabile, or "wonderful net," first identified in tuna by a French anatomist in 1830, is a complex system of veins and arter-ies that keeps a bluefin's blood away from its gills and within its muscles. While most fish evolved with circulatory systems in which blood flows alongside their spines and then branches out to muscles before returning along the same pathway, the blood of a bluefin tuna flows along its sides, just underneath the skin, before branching inward. A bluefin tuna's blood contains a high propor-tion of oxygen-carrying hemoglobin compared to the blood of most other fish in the ocean, and its speed partially comes from its warm muscles, which can contract faster than cold muscles. It is this heat that allows a tuna to catch, chase, and digest prey at incredible speed, while also enabling it to regulate its body temperature at nearly human temperatures, between 26 and 32 degrees Celsius, while living in waters between 3 and 30 degrees Celsius. Its warmer stomach also allows it to digest food faster. As long as a tuna manages to feed regularly, it can keep its body hot. But unlike a swordfish—another strong fish capable of fighting for a long time—the longer a tuna fights, the more it can overheat its own flesh, flooding it with lactic acid and resulting in a "spoiled" taste.

Then there's the matter of ice. Dragging a giant, ungutted fish alongside a boat for hours or leaving it lying on a ship's deck, which many Atlantic fishermen did with the giant fish, also spoiled the meat. Every good sushi buyer at Tsukiji knew that bluefin meat had to be cooled quickly in order to retain its quality. So Japanese buyers encouraged fishermen to bleed each fish and cut its head off as quickly as they could, and to install iceboxes big enough to hold three or more 400-pound bluefin at a time. Some fishermen harpooned the tuna with electrified spears, which some buyers believed resulted in a better product. Others encouraged fishermen to focus their efforts on landing fat, football-shaped fish that would bear the prized marbling that reached top-shelf prices in Tokyo.

Moon paid close attention to these developments in the market for tuna, as did his acolytes. In the weeks following the church's controversial purchase of the former convent, Moon decided a faster way to ingratiate his church into the still-hostile Gloucester fishing community was, if not through claims of peace and friendship, through fishermen's wallets. In the late 1970s, says Hokanson, three of the town's fish merchants had secretly organized to cap the price they would pay for bluefin at 75 cents per pound. That price was previously unheard of but, given the windfall the fish could land in the Tsukiji market, still far below what dealers could make selling top-grade tuna to discerning Japanese clients, even after paying for handling and airfare. After the church's purchase of the lobster pound, Moon directed its employees to start offering fishermen $1.50 per pound for tuna—double the previous price and what other dealers were offering. The other buyers howled, but within two weeks they, too, were paying the formerly inconceivable sum for bluefin. As quickly as fishermen could dream of it, that $1-per-pound ceiling for tuna had been broken. Once everyone had adjusted to the new price, Moon and his employees intervened for a final time: they would offer $3 per pound for fish they could sell abroad. Again, the other buyers fiercely complained, but within weeks, that was the standard going rate on the Gloucester docks. And it was still a fraction of the profits a Japanese sushi restaurant could earn selling the raw fish to diners craving the melt-in-your-mouth texture. It was a craving that would soon arrive in North America as well.

Starting in the 1900s, waves of Japanese families arriving in the United States would indelibly change the country's dining scene. Back then, most white American diners regarded uncooked fish as a novelty food, familiar mostly to those who had served overseas during the war or had eaten bluefin in Hawaii, where it was served as poke (Hawaiian for "cut crosswise into pieces") at luaus.

In 1926, a group of Japanese grocers in Southern California banded together to open Mutual Trading Company, an importer that supplied fancy rice, condiments, and niche products primarily to Japanese grocery stores. By 1964, Mutual Trading president Noritoshi Kanai had met and partnered with Harry Wolff Jr., a Jewish businessman from Chicago who had fallen in love with nigiri, raw fish served atop rice, during one of the duo's trips to Japan. Wolff encouraged Kanai to recruit a chef friend to open a sushi counter within Kawafuku, a Japanese restaurant in Los Angeles's Little Tokyo neighborhood. Its L-shaped, dark-paneled bar curved around seven oxblood-red diner seats, which faced a clear, ice-lined raw fish display with Japanese-style black-and-white textile banners and latticed screens above. Over the next four years, its *itamae* ("in front of the board") sushi master Shigeo Saito and his wife—his partner and only server, whose name never made it into the history books—made $30,000 in profit (more than $250,000 today). When locally caught tuna was available at a high-enough quality, Saito used it in his sushi rolls, a preparation made by rolling raw fish into dried-seaweed wrappers, or nori, along with sushi rice, vegetables, and garnishes, including sesame seeds and fish roes.

One of the restaurant's first competitors to serve raw fish on its menu was Tokyo Kaikan, a large complex initially designed in a Polynesian style, where, as recounted by Corson in *The Story of Sushi*, chef Ichiro Mashita first invented the California roll. It was apparently a seasonal lack of Pacific bluefin, which was available only in the summer months during its migration northward along Mexico's coast, that led to the roll's invention in the 1960s. When bluefin was in season, Mashita used its marbled belly, its toro, in nigiri preparations that quickly became popular with his regulars. When it wasn't in season, he and his assistant tried to recreate the bluefin belly's fatty texture, even experimenting with cuts of beef and chicken, until they landed on an unexpectedly vegetal substitute: creamy, California-grown avocados. Using sheets of nori as covering, Mashita reportedly rolled up a mixture of crab meat, avocado, and mayonnaise before slicing it, resulting in a dish that other chefs imitated and innovated upon across the continent. (Sasha Issenberg recounts a similar story in *The Sushi Economy*, with Kaikan's business head taking credit for the idea, while an Associated Press news article from the 1960s credits another chef, LA's Ken Seusa, for its creation. Meanwhile, Vancouver's Hidekazu Tojo emphatically insists he was the roll's creator, so the truth of who first rolled crab into sushi with avocado is likely to remain lost to time and memory.)

In 1963, writing in the *New York Times*, food critic Craig Claiborne feted the opening of two new Japanese restaurants where "New Yorkers seem to take to the raw fish dishes, sashimi and sushi, with almost the same enthusiasm they display for tempura." News accounts and reviews framed eating sushi as an experience for the culinarily adventurous and "delicious despite its appearance." By 1966, Claiborne found himself gobsmacked by the rise of Japanese restaurants in his city. In "New Yorkers Take to Tempura and Chopsticks with Gusto," he observes that "Americans for whom 'Chopsticks' was once a childish piano exercise now wield them with admirable expertise," with some dining on the raw fish "with a gusto once reserved for corn flakes." In 1970, Osho sushi bar launched outside Los Angeles's Little Tokyo neighborhood, catering to the tastes of fashionable movie stars. In 1977, *Esquire* declared the coastal trend official with the headline "Wake Up, Little Su-u-shi, Wake Up!" and by 1981 a posh sushi bar had opened in Manhattan's Harvard Club.

With this new market emerging and slowly spreading across the country, opportunity beckoned, particularly for men like Moon who had already positioned themselves at the whirling core of the tuna market. Throughout the late 1970s and early 1980s, Moon's early grandiose plans for the Unification Church to corner the burgeoning American sushi market quietly solidified, fueled by funding that came largely from his thousands of followers in Japan. Following his orders to vertically integrate his seafood empire, included owning everything from individual tuna to the fishing boats to companies that sold fish, Unification Church operatives started branching out into other business realms beyond fisheries across the United States while also trying to attract new converts. A 1976 rally at the Washington Monument drew around 50,000 people, a mixture of believers and gawkers. While US-based "mobile fundraising teams" like the one Hokanson encountered in Germany ate peanut butter sandwiches and crowded into shared apartments to save money, the church bought Manhattan's rundown art-deco-style New Yorker Hotel to turn into its new mission headquarters. By 1977, one church member reportedly told *National Fisherman* editor Tim Sullivan that the church benefited from being able to sell its bluefin directly in Tokyo without the usual costs of middlemen and selling fish to overseas retailers.

On April 16, 1980, soon after the 2,000-room New Yorker Hotel reopened after renovations, Moon addressed a 70-odd group of followers in its grand ballroom, as Takeshi Yashiro, a 30-year-old Japanese-born son of an Anglican bishop, watched in awe.

"You are the pioneers of the fishing business—the seafood business," Moon told them. "Go forward, pioneer the way and bring back prosperity." Their goal, as Yashiro understood it, was to draw on the seas, as Jesus once did, to fight world hunger, which they would do by selling fish door-to-door directly to consumers' homes, using a fleet of refrigerated vans. Moon gave each follower a crisp $100 bill and sent them across America.

On a public-relations level, Moon regarded most Americans' unfamiliarity with how to prepare raw fish as an advantage, not a curse. "Many Americans don't know how to prepare fish, so we do it, and all they have to do is eat it," he once said in a speech— although to be fair, in the same speech he also declared that he was testing recipes for a nutritious bread made from dry fish powder, which he claimed could be stored at room temperature in open air. "The best test will be giving it to children, who are very honest," he said. He was, in his own mind, a future food messiah. (The fish-powder bread and cookies never caught on and the equipment they bought for it kept breaking, so the project was eventually abandoned.)

As the Unification Church influence in seafood and shipbuilding grew, Moon found other ways to bring attention to the church, including holding a "world tuna tournament" in Gloucester, with plans to sponsor similar tournaments in four other locations. The top prize at each tournament would be $100,000, which Moon claimed would "lure tens of thousands of influential people." Moon, who declared that he personally wouldn't enter his own tournament out of a sense of fairness, delighted in the prospect that newspapers then would have to cover the superior design and skill of Unification Church–linked boats and bluefin fishermen such as Hokanson. "When the media finds out that my boats win the world tuna tournament, that will really be news," Moon declared to his followers. "They will wonder when this religious leader became such a fantastic boat designer." To the surprise of no one in Gloucester, Hokanson's boat won the first trophy. Ultimately, the tournament ran only twice, and only in Gloucester, in 1980 and 1981. Unification Church members won both times.

Between evangelizing and musing on how tuna fishing could feed the world, Moon even figured out a way to turn romantic matchmaking into a business proposition, which he publicly declared during a 1980 speech entitled "The Way of Tuna." By introducing and marrying American converts to Japanese-born church members, he suggested, his church could circumvent the country's new 200-mile-limit laws. "There is a great deal of diplomatic pressure to use these fishing grounds which are monopolized by the American government," he said. "However, we are not foreigners. The Japanese companies see that Reverend Moon has a base in the United States, and furthermore, has so many Japanese people working with him who are engaged to Americans. They realize that we are their only link to America. We are becoming the best bridge for Japan to trade in sea products."

Fishing together on the *New Hope*, Moon had long told Hokanson that he would eventually find him a good Korean wife. So one November, Hokanson received some good news: Moon had shown Hokanson's picture to a beautiful young woman named Aeryeon, and the two were to be wed. It took six months until Hokanson caught a glimpse of his new wife in a tiny photograph, a picture so small that he had to wedge it beneath the magnifying casing of his fishing vessel's radar console. Every time he looked at his radar, Aeryeon gazed silently back. Soon after, he headed to Korea to first get legally married there, where he met her family and presented her with a pearl ring. In Korea, his new wife's family gave him a traditional Korean outfit called a *hanbok*. One night, he wore the outfit while he entertained the family with his best off-key rendition of "If I Were a Rich Man" from *Fiddler on the Roof.*

In 1982, after Aeryeon had moved to America, she and Hokanson took a trip to New York City for an event that made headlines around the world. On July 1, 1982, they joined 2,074 other couples at Madison Square Garden for the largest mass wedding celebration in history. Atop the white-carpeted area, an ocean of men in identical blue suits stood beside their soon-to-be wives, who were clad in matching lace-and-satin dresses. Mendelssohn's "Wedding March" played as couples walked past Moon and his wife, who sprinkled blessings of water on them. Moon, who had been ordained that week, led the group through vows in Korean and the exchange of rings. At the end of the ceremony, Hokanson and his wife raised their arms together with the crowd of new couples and shouted *"Mansei!"*—"eternal victory" in Korean. From the aisles, thousands of friends and invited guests and family members watched, some bewildered with grief at the sudden choice their children had made: to join the bizarre foreign "cult," to marry strange men and women some had met only days before. Outside the arena's metal detectors, dozens of protesters, furious that Moon had apparently brainwashed their family members, demonstrated with signs that read "Let Our Children Go!" and "Hitler, Jim Jones, Moon." Within the venue, one woman from Memphis looked on as, somewhere in the gathered masses, her sister stood with a new husband. "That man has caused more heartache in my family and everybody else's than anybody I can think of," she told a newspaper reporter as she pointed at Moon on the red-carpeted dais. "My mother can't even talk about it without crying."

As his public reputation continued to grow and his church's ocean-affiliated businesses thrived, Moon's interest began drifting away from bluefin toward politics and a wider influence over media. It was around this time that his questionable business dealings started to catch up with him. Soon after he founded the conservative newspaper the *Washington Times,* a federal grand jury convicted the controversial leader for tax evasion. In 1984, the same year Moon went to prison, 83 small, never-launched Unification Church skiffs were seized in Virginia, where Samuel Barfield, Norfolk's tax assessor, planned to auction them off for nonpayment of taxes. He had determined the boats, with their 225-horsepower outboard motors, were commercial, not tax-exempt religious property. "When I go fishing I pray a lot, too," he said. "And I pay taxes."

The church's public fortunes were declining, but the bluefin industry continued to boom. By the late 1980s, Al's friend and fellow tagger Charlie Donilon had caught and sold a 705-pound giant bluefin for $13 per pound: he made nearly $10,000 selling that one fish. With profits like those, it was obvious to Donilon and Al, swapping stories on the docks, that without new regulations the bluefin tuna was toast. Plummeting numbers had recently caused the US government to put in tougher new restrictions on how many tuna could be caught; for the Unification Church, that meant they could legally catch only one big bluefin a week. As the booming 1980s hit their stride, the church's days of pulling unlimited tuna from the ocean had ended. But for Moon's purposes, the tuna had already served its role: the Unification Church's vertically integrated network of boatbuilders, fish processing plants, and distribution businesses had already spread across America, and there were plenty other fish in the sea.

CHAPTER SIX

RED GOLD

~~~~~~~~

Spain, 1980s

Say Istanbul and a huge fishery comes to mind
Like a rusty cobweb over the Bosphorus,
Or sprawling off the Marmara coast.
Forty tunnies toss in the fishery like forty millstones.
The tunny, after all, is the king of the sea:
You must shoot it in the eye with a rifle and fell it like a tree,
Then suddenly the face of the fishery gets bloodshot.
—BEDRI RAHMI EYÜBOĞLU, "THE SAGA OF ISTANBUL"

More than 30,000 years ago, the Strait of Gibraltar was a broad plain. Lapping several kilometers from the limestone cliffs that now tower above its blue, continent-splitting waters, sea levels were roughly 120 meters lower than those in modern times, a height difference about the size of the Great Pyramid of Giza. In spring, as they had for thousands of years before the earliest hominid evolved, bluefin tuna migrated from the cold, deep Atlantic inward toward the Mediterranean, drawn by instinct and ancient memories of spawning in the inland ocean's shallower, warmer currents. At the time, the African and European continents were a mere 10 kilometers apart, separated by two distinct, deep channels that had not yet merged, and wouldn't for thousands of years.

Throughout the fall and winter, huge schools of millions of bluefin prowled the chilly Atlantic Ocean, feasting on its bounty of fatty mackerel and herring, building fat stores and millions of eggs and spermatozoa that would help them complete their annual cycle. These ancient ancestors of Amelia navigated using a combination of light, scent, and possibly electromagnetism. Each had a translucent pinhole atop its forehead, called a pineal window, which channeled light down a cartilaginous stalk to the pineal organ. That organ allowed each fish to sense light, possibly even beams from the moon and stars. Just before dawn and just after dusk, the fish plunged away from the ocean's surface to recalibrate their internal compasses. By sensing light during the day and tracking the sun's progress around the earth, they followed cosmic patterns that accompanied their ancestors and would guide their children. They oriented themselves in relation to polarized light in the water, and used shifts in temperature, salinity, and the directions of the currents they swam with and against to find their way. Some of their bones contained trace amounts of the iron-based mineral magnetite, hardly surprising on a planet beset with electromagnetic waves—waves that could provide clues on where the tuna were and where they were heading.

Heading eastward, the outflowing ocean current was strong,

but so were they. In the open ocean they were kings, but in the narrowing bottleneck of the strait they were suddenly transformed into prey themselves, now pursued by pods of canny orca whales. It was a race some of them couldn't win, their fast, stiff bodies darting and cornered, diving and leaping out of the water. At least they had their speed. That speed was their defense, but could also be their downfall. Blinded by an instinct to escape, some fish rocketed onto the shallow beaches and shoals, where, as they had for countless seasons, small groups of Neanderthals waited, arms outstretched, for a gift from the sea.

Starting in 1989, the Gibraltar Museum supervised excavations of Gorham's Cave, part of a network of tunnels and chambers unearthed by colonial British engineers between 1782 and 1968, about an hour's drive from Cádiz, Spain. In 1907, Captain A. Gorham explored the high-ceilinged cave that would later bear his name. Tucking themselves into the Paleolithic caves, the modern researchers unearthed a trove of evidence of the Neanderthals who once sheltered there, covered by layers of sand gradually blown, grain by grain, into the cave by harsh easterly winds, drawn toward fires vented through the cave's 80-meter chimney. "Gorham's Cave is a time machine," evolutionary biologist Clive Finlayson told tuna writer and researcher Steven Adolf in his book *Tuna Wars*.

Throughout the 1990s, while exploring Gorham's Cave and other neighboring caves within a 28-hectare complex spanning the main ridge, researchers from around the world found charcoal, bone fragments, charred pine seeds, and what seemed to be blade fragments. They also found what they identified as "macroichthyofauna identifiable by tuna vertebrae of medium and large size"—or, in other words, evidence that both medium and large bluefin had been eaten within the caves. Paired with later-found evidence of fires and of tuna beachings caused by orca attacks in shallow waters, it signaled that even as the earliest modern humans spread across the globe, at least one hominid species already had figured out how to catch and consume tuna.

One of the researchers working in the field was a young professor at the Autonomous University of Madrid named Arturo Morales-Muñiz. In the mid-1990s Morales-Muñiz was widely referred to by Madrid's fishmongers as "the bone man." He visited their central fish market, Mercamadrid, every few weeks searching for the carcasses and bodies of their strangest creatures. Sometimes he'd buy a whole fish or a bagful, paying with coins he pulled from a battered leather change purse. Other times the fish were too large, like tuna or swordfish, so he'd settle for stripped, bloody skeletons. He loaded them into his trunk in leakproof containers scavenged from the market's garbage piles. His car stank, he knew, but it helped that he was "almost like a whale," he said, in that he had very little sense of smell.

In April 2022, I joined the tall, amiable Morales-Muñiz on a predawn visit to Mercamadrid, home of the second-largest fish market in the world after Tokyo's. Since 1982, cars have flowed past its entrance hours before the sun rises. Within its cavernous fish warehouse, thousands of people working for more than 100 companies operate forklifts, butcher fish, and sort a dazzling array of marine creatures by weight and size, quality, and when they'll spoil. Its aisles are closely packed with boxes of fish, cooler booths, and walk-in refrigerators with offices above.

Seven days a week, the market echoes with the shouts of fishmongers, some clad in blood- and ichor-stained aprons and ranging on a temperamental scale from furious to jolly. They're closely flanked and constantly approached by insistent salesmen, competitors gathering intel, and cooks in chefs' jackets looking for the day's fish specials. The day I visited, the sellers of fish were only men—men with beards and mustaches, bald men, old men, young men—who used whetstone-sharpened machetes, cleavers, and fine boning knives to separate bluefin flesh from bone and portion steaks. Their short, blunt fingernails scraped against the shells of shrimp and mussels as they weighed fish, shellfish, and a dizzying array of marine creatures on metal scales by the handful, the bucketful, the crateful.

Back in the early years, as Morales-Muñiz pursued his mission to gather as many animal skeletons as he could, he often found himself in bizarre and sometimes dangerous situations. What he was doing seemed insane, he knew, scavenging carcasses of "strange beasts" from the side of the road and harassing fishmongers for their strangest, most far-fetched and -flung fish. But it drove him crazy, how his country's archeologists seemed to worship only the relics and old walls left behind by the Romans and ancient Phoenicians, ignoring any bone that wasn't human. But if bluefin had indeed been the mortar of conquest and early Mediterranean civilizations, why hadn't his colleagues yet identified the fish's huge, arcing bones anywhere in the fossil record? For decades, historians and archeologists had insisted that the fish's calorie-rich body had fueled armies and provided early Europe with garum, a fish sauce that was one of its most expensive products. But if that was the case, why wasn't evidence of the fish being found on dig sites?

At first Morales-Muñiz claimed space in a huge, ancient storage building on the university's campus. He repurposed bottles and jars and food containers from his home, filling them with alcohol or formaldehyde preservatives and bones and scales. He invented a cataloging system for the jumbled remains, stringing boiled, dried spinal cords of giant mammals on heavy cords and rope and suspending them from a coatrack. The bones dangled vertically, each vertebra nestled within the next. Other bones and scraps of dried matter he tossed into plastic storage boxes.

Morales-Muñiz's father was a diplomat. In 1957, when Morales-Muñiz was five, his family moved to Morocco, and it was there that young Morales-Muñiz fell in love with the sea. Like bluefin scientist Molly Lutcavage, he admired Jacques Cousteau. He had watched Cousteau's *The Silent World*, entranced, and would later take the marine scientist's warning to heart: the sea was not inexhaustible.

When Morales-Muñiz was seven, in 1960, the 5.8 magnitude Agadir earthquake shook western Morocco, and about one-third of the city's population was killed. The night of the quake, the family slept in his father's Mercedes, and eventually relocated back to Spain until the family moved to Los Angeles, where his father worked as Spain's top liaison to the United States. There, Morales-Muñiz attended high school and took the SATs, although he failed them horribly the first time—his teachers gave him the test in Spanish, but he realized only too late that it was in the vernacular of nearby Mexico, not the language of his youth. It took a visit by his father to the then governor of California, Ronald Reagan, to get a second crack at the test, but this time in English. Morales-Muñiz aced it and got into a biology program at UCLA. After graduating, he got scholarships that took him to study in Germany, Norway, and Denmark.

When he first started working at the Autonomous University of Madrid, after returning to Spain in 1979, Morales-Muñiz kept his lab afloat by charging other departments money to write bone reports about the discoveries they made on their sites, using the earnings to fund his own research, buy equipment, and carve out support for graduate students. When he presented his first bluefin paper at a conference in Africa in 1987, a French researcher started yelling at Morales-Muñiz from the back of the room, telling him he didn't know what he was talking about and hadn't discovered anything new.

Morales-Muñiz entitled his second paper on bluefin "Where Are the Tunas? Ancient Iberian Fishing Industries from an Archaeozoological Perspective," and it was published in a tribute collection for his recently deceased friend and mentor Juliet Clutton-Brock. "One can get the impression that our knowledge on the origins, development and characteristics of these [tuna] industries must be extremely thorough," he prefaced. "In fact, the opposite holds." In the paper, he lambasted his country's archaeological community for essentially ignoring bones or biological fragments that weren't human. He concluded tuna bones hadn't been found because, after the flesh they had been attached to had been turned to garum, the bones were likely left to dry in the open sun, then ground up and turned into fish meal or fertilizer.

Widely regarded today as the father of Spanish zooarchaeology, the gregarious Morales-Muñiz is an internationally renowned expert in identifying animal bones discovered at ancient sites. He has also hosted numerous visiting graduate students and researchers from around the world, and visibly delights in the global nature of his profession. After our visit to Mercamadrid, he and I drove to his office for a tour. Entering the basement room—one he called departmental dibs on when his initial warehouse storage space was demolished—Morales-Muñiz flipped on the lights and we were greeted with the snarling face of a giant bearskin rug that once decorated his parents' apartment. On one side of the room, a giant table was taken up with plastic trays of tiny bones from a dig site halfway around the world, in Mozambique. "All in all, I think I have a little bit more than twenty thousand skeletons," he said. "I think the Smithsonian has more than I do."

On the room's other side, atop a rack of towering metal shelves, sat a large plastic box of tuna bones he got from a friend in Seville. Morales-Muñiz lifted the lid, warning me about the smell, and rifled through the yellowed pieces he used to identify ancient bluefin. Sometimes he mails small, brittle branchial arches, loops of dried fish gills, to fellow researchers so they can extract and sample DNA when they're working on their own archeological sites. It's amazing how far his profession has come, Morales-Muñiz acknowledged. "It's not so clear that the bloody spine from a tuna is cultural heritage," he said, describing his work to me later, as we drove along a Madrid freeway. Laughing, I disagreed, saying that I found it fascinating. "Well, you're a freak," he guffawed back. "And so am I!"

A FEW DAYS after my time with Morales-Muñiz, I hopped a cheap flight south and rented a car to explore the tuna trail on Spain's southern coast. One morning, before the heat of the afternoon, I

climbed a set of flower-lined steps to the Cave of the Orcas in Zahara de los Atunes, a town literally named for its bluefin, where a
bulbous cliff opening looked out over the cave's mouth to the Mediterranean far below. On the ocean's surface, a complex architecture
of nets held by huge, bright buoys stood out against the dark water,
visible even from far above atop my rocky perch. Since the late
1970s, countless tuna have been caught in branching networks of
Spanish setnets as the fish followed their natural tendency to head
in one direction until they find an exit from wherever it is they've
gotten themselves stuck. This is the start of the *almadraba*, a sta-
tionary system first developed thousands of years ago to trap and
kill one of earth's most mobile, hardest-to-catch fish species.

Some Spanish researchers harbor theories about this camel-
colored limestone cave, worn into the rock by wind and water, its
textures curved and branching like foamed bone innards. Its
rounded ceiling is streaked red-orange with iron residue, and
along its walls are paintings made by some of the earliest people
to inhabit this coast. Their artistic medium was a blend of crushed
iron dust and lard or spit painted onto the rock, including one
deerlike figure that has been dated to 20,000 years ago. Another
symbol, like a branching Y, is speculated by some archeologists to
represent an orca, alongside dots, arrows, and lines they suspect
represent an ancient sun-clock used to mark the seasons and passing of time until the orcas and their massive bluefin quarry would
arrive once again. Now the entrance to the cave is barred, held
closed by a heavy rusted chain and a small brass-and-steel lock.
Researchers, including acclaimed local archeologist Darío Bernal-
Casasola, think ancient locals could see the black-and-white predators from this cave.

From 1,000 years BCE, the Phoenicians pursued bluefin tuna. Their earliest traps consisted of palm branches impaled in the Mediterranean's sandy bottom that would channel the fish into shallows where they could be harpooned. Later improvements included a *tiro* system to target fish riding the currents rushing inward from the Atlantic Ocean toward the Mediterranean. The tiro technique used boats signaled by high stone watchtowers to entrap tuna with nets. Once notified by a flare from a nearby watchtower that tuna were approaching, five boats would head straight out from shore, dragging a net behind them to draw the schooling tuna to a halt. The boats then arced around the group, drawing the deep net into a giant U-shaped curve that entrapped them against the shoreline. Once the tuna stopped, another boat, this one with a tougher net, branched into another interior line of the "rainbow," with the two parallel curves now pinning as many fish as possible against the shoreline. Once both ends of the nets were anchored against the beach, men onshore hauled the nets in, closer and closer, until they eventually dragged the fish within spearing distance. In Andalusian Arabic the word *al-madraba* meant "place to hit and fight." This annual killing provided tuna meat for grilling and smoking and salting, while its guts and viscera—blood and organs and bones and heads and cartilage—were salted and fermented in giant pits. Workers packed the salted fish into amphorae, and early businessmen shipped it across early Europe. It fed armies, that tuna.

For millions of years, tuna have migrated annually back toward their ancient breeding grounds, so full of eggs and sperm their stomachs atrophy. Bluefin have no interest or need for food at this point, which means they can't be baited. Underwater, they follow currents and use them for navigation, cruising alongside prevailing marine highways to save energy. Any bluefin tuna running eastward into the Mediterrean along its northern shoreline from mid-April to mid-June off the Spanish southern coast of Cádiz since at least the 1400s was likely to encounter a wall of nets that drew it away from shore into the maw of the trap. Called *tonnara* in Italy, *madrague* in France, and *almadraba* in Spain, the systems all typically worked the same way. Fishermen used boat-pulled nets or complex chambers of nets held to the ocean bottom by heavy stones, and each time a bluefin reached another junction of the net system, it turned slightly, tilting its body slowly in a slightly different direction, first right, then left, then right again. These traps employed the tuna's natural tendency to swim in a circle when encountering an underwater obstacle. This slowly coaxed each giant fish toward the one-way channel of the net's mouth, where it entered a terminal pen from which it could not escape.

In modern times, the stairs to the Cave of the Orcas are overgrown, cracked, and crumbling. A white-walled mansion overlooks the inconspicuous rock, modern wealth displayed via huge glass windows and security cameras. At the end of the Second World War, a group of Nazis fled to this remote coast, building flower-laced villas with names like My Last Refuge. There is no sign marking the cave, only sandy footprints and worn rock corners from tourists ranging far off the beaten path, an occasional field trip of young anthropologists, or a late-night rabble of drunken teens. Beer bottles nestle in the bushes near the cave's mouth, detritus left behind like the rediscovered amphorae of previous centuries. We humans leave our evidence on the earth.

Deeper into the Mediterranean, on the western edge of the soccer ball kicked by the boot of Italy, a cave adorned with Paleolithic art was discovered in 1949 by a Florentine painter stumbling upon Levanzo's rocky beaches. Before then, local residents mostly hunted rabbits in the area, but the tourist explored farther and was stunned by what she found. Like the grassy plains near Gibraltar, the grotto was once part of a limestone plateau above the flatness below, where sometime between 50,000 and 10,000 BCE, early humans lived and worshipped, carving images of bulls and horse-like figures into the cave's stone walls. Thousands of years after that, other early artists mixed charcoal and animal fat to paint figures of four-limbed creatures, idols, and the distinctive fins and sickled tail of what appears to be a bluefin.

As far as historians and archeologists have been able to tell, the Mediterranean fishery for bluefin is probably the world's oldest intensively organized fishing industry. In ancient Greece and Rome, bluefin tuna were detailed in poems and stories, with Oppian of Cilicia describing "vernal trains" of the fish, the figures of which adorned goblets and cups used at feasts and in ceremonies. On one fourth-century BCE ceramic vase, a bare-chested merchant holds a giant machete above a tuna's beheaded corpse as he haggles with his robed and bearded customer. In the treatise *De Natura Animalium*, Roman author Claudius Aelianus wrote about fishing methods of Sicilian fishermen, including how, like the Phoenicians, fishermen used tall lookouts to signal to boats far below when they should open and close their nets to entrap the seasonally migrating school of tuna.

The image of the fish itself, in a physical representation of value, was even embossed on the region's earliest currency. One coin featured a Greek temple propped up by columns of vertical bluefin, while on another, a Greek coin traded in Cyzicus (now Turkey), a skinny hound crouches over a fat-bellied tuna.[1] On one silver Macedonian coin, a tuna rests at the bottom, topped by the scene of a lion (a historic symbol for fall) attacking a bull (a historic symbol for summer)—in a literal sense, says amateur historian Goran Pavlovic, the seasonal arrival of the tuna was foretold by the money itself. While warm-blooded mammals were commonly regarded as befitting the best sacrifices for the gods, only the finest bluefin tuna were sacrificed to Poseidon, their gushing, bright-red blood blanketing the altar in a tableau honoring his maritime kingdom.

In 350 BCE, Aristotle was the first scientist to record details of the bluefin's annual migration; he concluded it was making the trip to spawn in the Bosporus Strait. Three hundred and fifty years later, the Greek geographer Strabo similarly recorded the Phoenician pursuit of tuna past Gibraltar, as well as the establishment of a tuna-salting operation near Cádiz. Similar operations existed around the Mediterranean and were typically built near sea-salt harvesting businesses called *salinas*. In Baelo Claudia, Roman merchants salted bluefin tuna chunks, stripping and fermenting the fish's viscera into garum. The pricey, shelf-stable sauce tasted similar to modern-day Vietnamese *nuoc-mam* fish sauce. It was nutritious and protein-rich, containing minerals, B vitamins, and amino acids, including glutamic acid, a relative of flavor-enhancing MSG (monosodium glutamate).[2] Roman author and naturalist Pliny the Elder also reported nurses feeding young children ground-up tuna liver to help them grow.

These boat-based fisheries evolved into trap-based operations that, in turn, eventually evolved into one of the world's largest and longest-lasting bluefin fisheries, which was first conducted in Sicily when the island was under Arab governance. In Cádiz, the 1st Duke of Medina Sidonia, Juan Alonso de Guzmán, earned the nickname "dios de los atunes" (god of tuna) for his pioneering interest in the region's almadraba fishery, to which he had been given exclusive rights by King John II in 1445. He conceived of and ordered the building of the Chanca o Castillo de las Almadrabas complex to house the region's fishery businesses and a consolidated tuna market. Once a year, Guzmán would descend upon the castle, flanked by his noblemen and occasionally even King John II and his wife, Isabella.

But the boom wouldn't last. In 1757, a Benedictine monk wrote a letter to Guzmán's descendant, the newest duke, observing a decline in catches in the almadraba traps, which had plunged from a high of 58,000 individual fish caught per fishing season two centuries before. "In past centuries the tunas caught in the traps were almost infinite," he wrote, but within twenty years of those record catches the traps were catching only a mere 5,000 fish annually. For hundreds of years, fishermen cursed the orcas that hunted the giant bluefin they themselves felt entitled to, damning the sharp-toothed whales for their speed and strength and blaming them for declines in their catches. In 1774, Guzmán's staff even developed an elaborate plan to build life-size wooden orca whale models that could herd the increasingly sparse tuna schools toward duchy-controlled nets. But scientists studying the ancient catches have more recently concluded that the declines actually may have been linked to low temperatures during a period of cooling in the

North Atlantic dubbed the "Little Ice Age," which may have affected the bluefin's spawning-driven migration patterns.

A FEW YEARS before the end of the Second World War, a Spaniard named Diego Crespo Sevilla sensed opportunity. At the time, his country held a colonial protectorate in northern Morocco, and bluefin tuna, Crespo knew, ran along both sides of the Mediterranean. He moved his family there and eventually set up an almadraba-style setnet system near the former Phoenician town of Lixus. Centuries prior, he learned, bluefin had been caught there by the thousands. When the Spanish protectorate in the country ended in 1956, the family returned to Spain, but bluefin— and the profits they generated—stayed in their blood. So when the state-run consortium closed and the state monopoly on the fishery fizzled in 1971, the Crespo family jumped to fill the void. And they weren't the only ones. Another family, the Ramírezes, pulled strings within the federal government to lay claim to the tuna trap off Barbate, about 15 kilometers north of Zahara. The families hated each other but ultimately reached an uneasy brokered compromise: to the Crespos went the Zahara trap, to their rivals the Barbate location. Soon after, the Crespo family encamped on a large, white-walled complex that was once the bluefin consortium's headquarters, including a processing plant and salting facility, along with a cafeteria and small hospital for tending wounds inevitably inflicted by fish and fishing gear.

Despite the potential the two families had identified, those early years trapping bluefin floundered, and in Zahara the company barely broke even as prices for canned bluefin slumped. It was then that the Japanese arrived in southern Europe, just as they had in Rhode Island and Wedgeport, looking for bluefin.

As Japanese businesses hunted for new sources of bluefin tuna throughout the early 1970s, they spread their agents across the world, identifying new potential sources like those in eastern Canada and the United States, but were also fully aware of the ancient history of bluefin and those formerly flourishing ports along the Mediterranean. In 1978 the first sale of bluefin from Cádiz was made to Japan, and by the 1980s it wasn't unusual to see a state-of-the-art Japanese reefer floating off Barbate. Fishermen struggled with the culture change their new customers' high standards for freshness imposed: accustomed to a well-earned cigarette or beer after the catch was landed, the fishermen instead had to rush to land, process, and freeze the fish. The prices they earned from these buyers were worth the extra bother, and the almadraba appeared to have been a good bet after all. "Without Japan the almadraba would have disappeared," Diego Crespo, son of the original Diego, told *Tuna Wars* author Steven Adolf.

Working from a stand in Madrid's giant commercial fish market, Ricardo Robles, a veteran fish salesman, recalled how 40 years ago he'd never see a bluefin tuna passing through the market because it all went straight to Japan. In the 1990s, as production and demand increased, local chefs started asking for the fish, he said, and the seafood companies began to reserve some fish for a growing Spanish market. Soon after, Robles found himself selling bluefin steaks destined for grilling on the *plancha*. Almost overnight, it seemed, he could sell the fish for $20 per pound for use in tartare or sashimi, and demand for the fish blew his mind.

Along with sky-high prices arrived a wave of corruption and illegal trade. Even a global crash of bluefin in the 1960s hadn't stopped many big boats from operating in the Mediterranean, and in the years following, millions of small bluefin had been scooped from the eastern Atlantic, wiping out decades' worth of the species' future breeding contingent. With very little oversight from federal fisheries departments and no rules yet enacted by the International Commission for the Conservation of Atlantic Tunas (ICCAT), it remained laughably easy for big, internationally flagged boats to catch bluefin and transport the fish directly to Japan without reporting the landings to government or paying taxes. That, paired with a decades-long collapse in local employment, left Spain's once fishing-dependent region of Andalusia with only about 10,000 local fishermen plying its nearby waters. For a culture rooted in the fishery, it devastated families, many of whom turned to smuggling and drug running across the Mediterranean and southern Europe to support themselves.

They were strange, uneasy times, as families found themselves split between ancient culture and profits, between catching and selling as many bluefin as they could to a foreign nation while lamenting the loss of those same fish in their own waters. Globalization and prices the market was willing to pay for tuna had, once again, entrapped Barbate and its neighboring communities in golden handcuffs. That newfound wealth drew a new influx of tourists, as well as the interest of criminal networks who saw the fish's spiking value as an opportunity to smuggle fish across international borders. Human appetites paired with near-unfettered access to the ocean and its creatures had set tuna on the precipice. Now it was only a matter of time to see how far the species could fall.

CHAPTER SEVEN

KINGS OF THE OCEAN

~~~~~~~~~~~~~~~~~~~~~~~~

Al, 1980s

And the king said, Bring me a sword. And they brought a sword
before the king. And the king said, Divide the living child in two,
and give half to the one, and half to the other.

—1 KINGS 3:24–25, KING JAMES BIBLE

To be a bluefin tuna during a tuna tournament in 1980s New England was to be besieged: by scent, by sound, by sight. Floating on the waves, bloody fish pieces smeared the water with their seeping juices. A bluefin, using its brain's two large olfactory bulbs, can sense even the tiniest trace amounts of oily residue from a single intact fish blithely swimming in its vicinity. Those chum pieces exploded like fireworks in the bluefins' brains, drawing them out of their orderly formations into a feeding frenzy. Among those gory hunks floated baited hooks strung with herring with their backbones removed so they moved more naturally underwater, with weights inserted into their bellies and then stitched closed. The sneaky technique made the baitfish appear alive and healthy, so when a tuna lunged, the hook protruding from the herring's belly would lodge in its mouth. The fight was then on.

In the fall of 1980, standing in the gravel weigh-in area at Point Judith, Rhode Island's first Masters Invitational Tuna Tournament, Daryl Anderson, then in her late 40s, basked in the bright, warm light of Al's adoration. Earlier that day, despite her tendency toward seasickness, she had agreed to join Al's crew for their third day of tournament fishing. Out of 83 anglers fishing from 26 boats, only 3 that year were women, and Al was proud that his wife was one of them.

The first two days of the competition, which were predicted to be rough, had dawned crisp and clear. The three fishermen from Connecticut who paid Al to take them out for the tournament bragged about the fish they caught over those days: the first, 408 pounds, the second even larger at 493 pounds. Like all fish caught during the tournament, Al trundled those fish back to the docks at the Port of Galilee so they could be sold into the hot Japan-destined bluefin market, with fishermen and the event's organizers splitting the profits. After their first two landings, Al, who always kept an eye on the competition, privately hoped that the *Prowler* might win the trophy. Al loved a good trophy.

On the third day, Daryl and the group climbed aboard Al's eight-meter Bonito bass-fishing boat and headed to the tuna grounds as the winds began to rise. The group decided, by consensus, that if they caught a tuna that day, Daryl would take the rod. But as the boat climbed into the surf for more than an hour, the waves grew larger, rolling hummocks of water that left Daryl feeling nauseated and disoriented. They were now 40 kilometers off the coast, and Al could see other boats around them hooking up to tuna. Daryl, fighting the urge to vomit, tried to sleep off the worst of her seasickness while Al used a rod and reel to jig a few silver-quick whiting from the ocean bottom. He hooked one onto a leader and trailed it near the bottom, when finally—finally!—on his fish-finder he glimpsed a tuna cruising near the seafloor.

Daryl awoke to shouts: a tuna had taken the whiting. Line screamed off the heavy rod as the men on the deck scrambled to pull in the other rods and get the boat off anchor. In the first seconds of a tuna hit, the easiest way to lose the fish is to tangle with another line: the new extruded monofilament they had started using in the 1970s was infinitely better than linen or piano wire but was designed to be pulled taut, not struck by another line from the side. There are few things as disheartening to an angler, as I know from personal experience, as the feeling of a fish on the line, heavy, present, alive, followed by a jolt and then nothing, a feeling of negative space as your shredded line floats free on the waves. But Al and the three men got all the lines in, hollering as they went. As Daryl pulled herself upright, roused by the commotion, a look of dawning horror creeped across her face. It was her turn to land a fish, but she still wanted to throw up and could barely move.

On Al's direction, two of the Connecticut men stepped toward Daryl and carefully lifted her under the arms and set her into the boat's fighting chair. Daryl, who had never caught a giant tuna before, hazily said, "I don't think I can do this," as her husband slipped a pair of wet cotton gloves onto her hands to help protect them from blisters. The tuna had rocketed quickly away from the fleet, which meant they didn't have to worry about tangling lines with other boats. Once Daryl was in the chair, Al notified the judges by radio that they were chasing a fish, and now it was time to get back some line. Daryl, not used to being strapped into the elaborate, hinged fighting chair, gasped every time the fish yanked the line, tipping her face-first toward the water. She knew the chair's heavy rope straps were designed to hold her and the rod tightly to the chair, but that didn't help much. Drag pressure whined on the reel—45 pounds, then 50 pounds—as the tuna raced below the surface, trying to escape. Suddenly, she yelled for Al: she needed a bucket. Even the excitement of a tuna on the line couldn't extinguish her seasickness. "It's not a good time to get seasick, honey," Al said. Daryl replied with a profanity. But Al's interjection had done the trick. She was going to land that fish.

Well away from the fleet now, the Andersons worked in tandem. From the cockpit, Al powered the boat in the same direction the fish was running to help Daryl take up slack, while she gradually pulled line, which was marked at 15-meter intervals, back onto the reel. After nearly an hour the fish surfaced, and Daryl, her arms numb with exhaustion, knew she was close. She cranked on the line as Al roared the *Prowler* onward. Then the boat's steering quit: in a stroke of bad luck, an internal oil leak had sprung and the boat was rudderless. But Al always planned for everything. He poured his spare oil into the engine and finally, finally, the fish was swimming in desperate last-gasp circles that meant it had nearly given up. Al explained to Daryl what the closing moments of catching the fish would look like and he climbed down from the boat's tower to operate the boat from deck level. One man "leadered" the fish, controlling its front from the heavy wire leader attached to the hook in its mouth, while another hooked the fish with a gaff as Al slipped a rope around its tail.

The marine radio buzzed with congratulations for Daryl as the boat headed back to the fleet. The men on the boat hugged her and kissed her on the cheeks as she slumped back into the fighting chair, her forearms tight from working the reel, her legs wobbly from the struggle. Al measured the fish with a tape measure and crunched some numbers on a piece of paper on his console. It looked as if the fish was only about 300 pounds, which meant it fell short of the 310-pound eligible minimum for the tournament. But they wouldn't know until they put it on the scale.

As the fog closed in, the fish Daryl landed was hung and weighed: 307 pounds, only 3 pounds short of the cutoff. Their team would take second place. Standing on the rocky pier, a crowd of onlookers packed behind them, Daryl grinned toward the tournament camera, her right hand resting gently on the tuna's small belly fin, her left holding the brass-reeled rod. Al, who typically saved his smile for the camera, stood on the other side of the fish, one hand on his hip, wearing a crumpled bucket hat and his big, slightly tinted eyeglasses. The beam of his affection pointed toward Daryl—the woman who spent her mornings taking detailed accounts of his shipboard records, who made him lunch every day, who had given him children he had long assumed he'd never have. She was perfect for him, and she had also given him the confidence to make the next big step he knew, in his heart, he had to make: he was going quit his teaching job and try fishing for real.

By 1980, Al had built up $50,000 in his savings account—money he meticulously put aside by reusing string, selling any Japan-quality fish he happened to catch, and working every day of the week the ocean allowed. The day he hit that amount, he arrived home with a bottle of champagne. He and Daryl opened the bottle and savored its bubbles. After finishing the bottle, Al pocketed the cork and took it down to his basement desk. He pulled out a pen and scrawled "$50K" on its bottom. It was an amount he had long imagined but found hard to believe was real.

Al had been teaching for 20 years, long enough that he could retire with a small pension, which would help cover the cost of a new *Prowler*—he'd need a new boat, after all, to lure clients aboard. After some asking around, Al settled on buying a 10-meter diesel sportfishing boat with a high rooftop steering tower built by Connecticut builder Peter Legnos. Al dipped into those hard-earned savings to buy the boat and set up as much of the boat's interior as he could himself to save money. Without his teaching salary, it was tough, but Al hoped that someday he could make enough money fishing to cover the family's bills.

One of Al's first purchases for his new full-time business was white collared shirts with pockets and the boat's name embroidered on the chest. He took his inspiration from a well-to-do plumbing entrepreneur who ran a charter out of Snug Harbor, Charlie Bouchard, a big fish tagger and respected captain. With those crisp shirts and a matching cap, Al wanted to look like he ran a professional operation.

When Daryl caught her very first bluefin in fall 1980, US-flagged vessels were landing around 234,000 tonnes of bluefin and smaller tuna species every year, drawn primarily from American waters. Those fish were worth nearly $300 million at the dock, but once processed and canned were worth four times as much. Most canned tuna at the time were Pacific bluefin, *Thunnus orientalis,* caught off the coast of California and farther afield off the US Pacific coast. Only one-quarter of US-caught bluefin was caught in the Atlantic. Yet it was debate over those Atlantic fish that, in the early 1980s, set the entire global bluefin fishery on a collision course with its controversial, billion-dollar destiny.

As countries tussled over the high-value fish and increasingly patrolled and defended their self-declared territorial waters, the two intergovernmental groups established to mediate exactly these kinds of state-level conflicts, ICCAT and the United Nations, were listening to reports from scientists, including Frank Mather, who said the bluefin remained in serious trouble. The only reason it was still possible to catch bluefin off New England, he believed, was thanks to the restraint shown years before by the purse seining industry and new ICCAT limits on catches of small tuna, or, to his mind, the species' prospects.

Meanwhile, most bluefin fishermen scorned Mather's dire predictions. "I only spent half my life at sea, and there's Frank Mather sitting in his office, drinking cocktails and reading letters from fishermen in Sicily," Massachusetts fisherman Herb Randall told *Sports Illustrated* in 1974. "Sure Europe's tuna are in tough shape, but why pick on us? We've got a healthy fishery." ICCAT's member countries were going to have to make a choice, and what they decided on would shape the fate of bluefin for decades to come.

In the late 1970s, some members of ICCAT's scientific advisory committee started bandying about the idea of dividing the Atlantic bluefin population into two distinct "stocks."[1] At the time, scientists believed that Atlantic bluefin spawned and migrated back to two discrete breeding areas, one in the western Mediterranean and the other in the Gulf of Mexico, although they were the same genetic species. "The North Atlantic contains two main populations, which are more or less separate," wrote one Canadian scientist in an early report. "One lives on the European side and one on the American side." This was the conventional wisdom that Frank Mather's data sets, which were built using recaptured tags including Al's throughout the 1970s, had increasingly disproved. That data clearly showed that some bluefin continued to regularly cross the Atlantic Ocean in search of prey.

Ever since Al started tagging bluefin tuna for Mather more than a decade earlier, the single-minded fisherman had spent hundreds of hours building gear and perfecting systems for catching "footballs" that schooled on the outskirts of his favorite fishing spot off Block Island. In 1976, he tagged and released a two-year-old, 18-pound bluefin caught by one of his stepdaughters aboard the *Prowler*; two years later the same fish was recaught by a commercial fisherman in Europe's Bay of Biscay. It was Al's first transatlantic "recapture," as the recovered tags were known, and was a recapture for which he had waited the better part of his career. He had no idea that the fish off Rhode Island were known to travel so far and at first he thought the report was a mistake, but Mather wasn't surprised. The recovered tag did, however, help reinforce the scientist's belief that if bluefin were going to be protected, they needed to be protected at an international level.

In May 1980, Mather drove on last-minute notice to Washington, DC, to give testimony to the US Congress's Committee on Merchant Marine and Fisheries, which was considering the continued funding of the bill paying for the country's enforcement of ICCAT regulations into the mid-1980s. At those hearings, commercial representative Dave Fyrberg decried the size limits on bluefin that had been imposed during the 1970s by "doomsday sellers," including Mather. The same year that American fishermen had been cut off at the knees by catch limits, he pointed out, Japanese longliners had caught 15,000 giant tuna in the western Atlantic. Furious with the ICCAT restrictions and Japan's apparent flouting of its limits, some fishermen and their representatives were lobbying hard to pull the United States out of the ICCAT agreement entirely. "We are allowing foreign countries to walk away with our natural resources," one angry fisherman told a US congressional committee in 1981. (A marine policy student and former bluefin hand-gear fisherman named Jack Devnew included the comment in his 1983 master's thesis for the University of Delaware. "No one can be considered an unimpeachable source in a controversy such as this where all parties seem to have a vested interest or an axe to grind," he concluded.)

Mather, after reading from his hurriedly prepared statement at the committee's wooden table, balked at the idea of the United States leaving ICCAT and walking away from what few regulations the body had imposed on bluefin. "A little exercise of human intelligence, for a change, by fishermen and regulators, [has] given us a chance to save the Atlantic bluefin tuna," he told the committee. "If we resume the slaughter of small fish to the extent that practically none of them escape to become spawners, we will be in bad shape." During the hearings, Al's recaptured Bay of Biscay fish was specifically mentioned by another commercial representative: that single tag, and dozens of others caught in the 1960s and 1970s, had provided the first glimmers of solid proof that the bluefin's life cycle and migration patterns were far more complex than anyone knew.

The growing number of tag recoveries from Al, along with those of other fishermen he corresponded with, convinced Mather and many of his peers that bluefin from both spawning grounds met and mingled across the Atlantic, running as far north as Norway and as far south as the Canary Islands. But for some ICCAT scientists, the number of fish that ended up being caught on the "wrong" coast was incidental and considered to be so few as to be inconsequential. It was known that the fish mixed, but at ICCAT, an organization that had been formed with the goal of providing MSY-based quota allotments so that fish could be pulled from the ocean at the ecosystem's absolute upper limit, the prospect that they would have to incorporate mixed stocks and the associated uncertainty into their equations also seemed frustratingly inconvenient.

For two-stock advocates, their theory facilitated a convenient mathematical-modeling shortcut that made it easier to calculate and enforce limits on how many total Atlantic bluefin could be caught on either side of that 45° line separating North American bluefin from European bluefin. Yet that statement of scientific "fact," as it was then understood, was never intended to provide scientific certainty but instead to help address a politically fraught international conflict. Many of the American two-stock proponents hoped the policy would entrench the control of their country over its own territorial waters.[2] Along with globalization had arrived greed, and politicians and policymakers had proved themselves, time and again, to be easily swayed by tax dollars, reelections, and lobbying perks that came with prioritizing extraction regardless of the environmental consequences—first of oil, but also of other natural resources like fish and timber and coal and metal ores. The situation had all the makings of a climate and ecological disaster, of which the bluefin's predicament was a tiny slice.

As an apex predator, bluefin tuna fill an irreplaceable niche in the ocean's ecosystem. Every day the fish can consume 8 to 10 percent of its body weight in other marine species, consumption that, when multiplied over the fish's entire range and population, plays a huge role in influencing and controlling populations of the fish and invertebrates it eats. Without bluefin tuna, a phenomenon dubbed "trophic cascade" would occur as the butterfly-wing effects of the loss rumbled throughout its broad former range. Like deer overrunning a forest in the absence of wolves, the bluefin's prey species would tear through their own available resources. (This type of imbalance occurred after the Newfoundland cod collapse in the 1980s and 1990s, as the absence of cod resulted in skyrocketing numbers of benthic invertebrates such as snow crab and northern shrimp, as well as small tunas and seals that fed on them.) The cascade would eventually force fishing communities to start "fishing down the food web," a term coined by University of British Columbia marine biologist Daniel Pauly, as they start targeting smaller species of fish in lieu of larger ones like bluefin and cod. (Pauly also coined the influential term "shifting baseline syndrome" to describe how we have lost our sense of how large, diverse, and plentiful the ocean's creatures once were and how steeply their numbers have fallen over time.)

A disappearance of bluefin would crash through the world's oceans with unforeseen consequences that are hard to predict, although one 2012 modeling simulation study of bluefin aquaculture found that numbers of bluefin in an ecosystem can deeply influence the health of smaller creatures including swordfish, mackerel, Norway lobster, and bonito tuna. Any damage to those creatures would then ricochet down to even smaller species like herring, squid, and jellyfish, as well as to microscopic creatures like diatoms and phytoplankton. This hollowing out along the food chain could eventually damage the carbon uptake of marine plants, potentially accelerating erosion and damage to coastal communities from storm surges and flooding, among other anticipated changes caused by global warming.

US modeling experiments done throughout the 1970s had already shown the scientific community and policymakers a new way to avoid that worst-case scenario. Atlantic bluefin tuna could be protected, and the solution was simple: end overfishing. To do so, either commercial catches on both sides of the Atlantic had to be reduced—something European countries staunchly opposed—or European levels could stay where they were and the western Atlantic catches by Canada, the United States, and Japan could be dialed back close to zero. Facing the winds of strong public pressure already mounting to reduce catches of bluefin, and with the Europeans unwilling to stop fishing bluefin, both Canadian and US policymakers found themselves in a tough spot.

In his thesis, Devnew highlighted the work of another researcher, Arild Underdal, who put forward his "law of the least ambitious" hypothesis. As summarized by Devnew, Underdal posited that in cases where international consensus is necessary to make a decision, collective action would always be limited to "those measure[s] acceptable to the least enthusiastic party." In this case, the European countries weren't ready to budge. With bluefin stocks clearly in trouble, that outcome meant catastrophe for the fish if nothing changed. And losing tuna, which play a key role in ocean ecosystems across the planet, meant more than just losing a glorious species. It meant upsetting, in a fundamental way, a delicate ecological balance developed over millions of years. It seemed far-fetched to many that a global appetite for sushi could wipe that out in mere decades, but that was indeed the stage that had been set.

In 1980, ICCAT's scientific committee released an "alarming" analysis, said Devnew: that if fishing levels stayed the same, numbers of adult tuna would "decline considerably by 1985." The following year, in 1981, the United States' own Southeast Fisheries Science Center published a troubling report that concluded that medium-sized bluefin, which were the future of the breeding-age fish, had decreased in the western Atlantic by one-third over two decades. By ICCAT's next annual scientific meeting in Madrid, the scientific committee had already reached emphatic consensus: the western Atlantic's bluefin population could simply not support the number of fish being pulled from the ocean, and catches of juvenile bluefin—and possibly even adults as well—had to be reduced to "as near zero as feasible." Without intervention, there wouldn't be enough small, medium, or giant fish to sustain the population, and a crash was imminent.

On many issues the bluefin delegations from the United States and Canada stayed entrenched at loggerheads. The Canadians had long complained about US reluctance to curtail catches, and according to Devnew, the American delegation had long resisted seeing and addressing how serious the bluefin situation had become, even though "the scientific data itself ha[d] always been there." So it was especially shocking when America's lead negotiators made their sudden, drastic proposal: the adoption of the two-stock dividing line and a moratorium on the fishing of all western bluefin, which a handful of recreational fishers and environmentalists had demanded the year before in front of Congress.

Unbeknownst to the other delegates, internal US discussions had concluded that managing bluefin as one population gave them less control over the bluefin in waters they wanted to claim as their patrimonial property. So the US delegates decided to throw the country's considerable weight behind convincing the international community to adopt a two-stock theory. Not because it was scientifically correct or proved, but because it was in the country's best political interests and arguably the only thing that could save bluefin in the Atlantic, given European countries' refusal to reduce their catches.

Using a two-stock model, the new US proposal called for a 565-tonne fishing-allotment quota for "ongoing scientific monitoring" for all countries currently fishing in the western Atlantic: Canada, the United States, and Japan. At the time, even the Unites States' own advisory committee and its state-level fisheries agencies hadn't been informed of the dramatic shift in position. It took three hours-long sessions, including two recesses, for the gathered countries to hammer out the wording for a proposal everyone was willing to consider.

Negotiators for Canada and Japan seethed, as did fishermen along the entire eastern and Gulf coastlines, as soon as they heard the news. As far as the Japanese were concerned, the moratorium was being, as Devnew put it, "rammed down their throats." News reports from Japanese spokesmen said that their country's delegation was "shocked" and "in trauma" regarding the proposal. "Sources present at the ICCAT meeting reported that the Japanese delegations appeared confused," recounted Devnew, "and resorted to calling Tokyo for instructions."

Canada aggressively negotiated to increase the total western Atlantic scientific quota to 800 tonnes, a change the commission agreed to incorporate into the US plan. And exactly where they would draw the line didn't ruffle many feathers on either side of the Atlantic, particularly given the fact that most of those delegates around the table knew the line itself was arbitrary. According to one attendee, the 45° line, which was between the two continents, was merely a "convenient fiction" of political expediency, designed to help the US delegation wrest some control from the unwieldy international community: the exact longitude of the line was not a detail on which delegates were willing to spend further political capital. "The two-stock theory was a political decision," Dick Stone, a former fisheries service scientist who later consulted for the Recreational Fisherman's Alliance, told the *New York Times*. "We couldn't get the Europeans to go along with strict conservation measures so we split the stock." The predicament echoed the biblical story of Solomon, in which two women fought over a baby, insisting he was theirs. When the ancient king declared he would cut the baby in half, the child's mother relinquished her claim to save the child's life. In the absence of action, it seemed as if the United States had decided to save the baby.

The western moratorium was approved at ICCAT's annual meeting, held in November 1981 in the Canary Islands. Japan was the obvious loser, but back in the United States, commercial bluefin fishermen and state regulators exploded, furious at the news. One company even sued the federal government, requesting a temporary restraining order against the implementation of the ICCAT policy. Boston seafood dealer Gerry Abrams, who estimated he was set to lose around half a million dollars a year (more than $1.5 million today) because of the moratorium, convened a meeting of bluefin fishermen at the New England Aquarium. The East Coast Tuna Association (ECTA) was born at this meeting, with the explicit goal of fighting the ICCAT-imposed limits. It raised more than $15,000 at that first meeting and used the money to hire a fisheries biologist and mathematician to review the government's data, which Abrams referred to as "bad science."

Within a few months, the US Atlantic bluefin purse seine industry ground to a halt. Federal regulators had declared a fishing moratorium that lasted two years. In an article published in a law review, a state of Massachusetts lawyer blasted ICCAT's management of bluefin as having caused "inequitable harm" to the region's commercial fishers. Roger Hillhouse and his business colleagues were forced to dock their fleet. The decision to close the bluefin seine fishery, that state's director of marine fisheries concluded, served no beneficial purpose, except to "fragment and alienate" the industry, even though it was clear bluefin stocks were in trouble. In 1980 and 1981, before the new management quotas came into effect, US bluefin fishermen didn't even catch the full quota they had been allocated. This was a sign to scientists like Mather that there simply weren't enough fish in the water to fulfill the relentless demand and that quotas should be slashed. "Fisheries management," Devnew wrote near the end of his thesis, "is more the management of people than it is the management of fish." As fishermen, environmentalists, and regulators were discovering, there were few comfortable détentes in an industry that could make or break fortunes.

THAT FIRST YEAR of the US moratorium on Atlantic bluefin, during the steamy summer of 1982, 13-year-old Steve Tombs grabbed the rubber-gripped handlebars of his secondhand yellow bike, yanked it upright, and jumped onto the seat. He pedaled furiously down a small hill, feet churning, trying to build enough speed to make it up the next rocky rise. His family's Snug Harbor cottage sat on a lane leading up to a slightly bigger lane, which itself led only two ways: back to town or down to the harbor. He picked his course and headed for the ocean: he wanted to make sure he was at the marina in time to greet the salt-scaled tuna fishing boats returning from a day on the water.

In his early teens, Tombs spent summers in Narragansett, Rhode Island, in a gray wooden summer cottage his family had built themselves over two weeks during their son's spring break. As he grew up, Tombs passed every afternoon he could watching the deep-keeled tuna boats arrive at the harbor. Using winches and machinery, fishermen hung enormous giant bluefin from a board painted with the marina's name and phone number; they nailed dried-out tuna tail fins, useless to Japanese tuna buyers, to the weathered dock posts where they docked their boats. They were gory, weaponlike trophies, proof of triumph. This was the height of the 1980s tuna price spike, when it wasn't unusual to have a half-dozen giant fish, often 400, 500, or 600 pounds, regularly hung up on the same day, one after the other. In previous decades, the giants might hang there for hours before being bought in bulk and sold as cat food, but those days—especially considering Japanese buyers' demands for freshness—they rarely stayed there long. Any day it wasn't pouring rain, Tombs ditched his bike, found a spot to sit, and simply watched, sometimes buying a soda or an ice cream. On Saturday afternoons, cars and people and trucks and trollies filled the marina. People gawked, embarked, disembarked; bought fuel and snacks; sold fish and bait. He spent his summer this way, and spent the winter dreaming of fishing.

The next summer, mustering up his courage, Tombs walked out on the flaking dock propped on the rocky shore and approached a familiar boat. Al Anderson, for all his fearsome reputation on the docks, seemed happy enough to answer an hour or so of questions while hosing off the *Prowler*. After a few weeks of hanging around when Al docked, the bashful 14-year-old boy wrote him a letter asking to volunteer as a mate: "Dear Al," it went. "You don't know me, but I'm the kid standing around on the dock watching you when you come in." Al couldn't bring himself to say no.

Starting every summer day in 1983, Tombs walked the sloping shoulder to Snug Harbor Marina well before the sun was up, taking care not to get hit by half-awake fishermen driving their trucks through the dark. Working at Anderson's side, Tombs fell in love with tuna fishing, which Al continued running under the one-fish limit because he could offer tuna-tagging charters. The next summer, Al even started paying Steve; the teen also made decent money from tips slipped to him by professional men, those lawyers and doctors and accountants who were chuffed at the number of fish they had caught. A discolored photo from that time shows Tombs, wearing belted khakis and his white mate's shirt, his name in cursive writing on the chest, standing beside a hanging tuna. A black, silver, white, and gold tuna rod rests in his right hand, its golden reel a tiny winch. His left hand is lifted nearly straight up in the air, clutching the bottom fin of a gaping silver tuna. Behind the hanging fish, three disembodied tuna heads caught by other boats lay wetly on the dock, each wider than his own body, the harbor's blue-gray water and faded sky filling the frame.

During those years, as Tombs worked alongside the man he called Cappy, the bluefin moratorium stranded dozens of American commercial fishing boats at their ports up and down the eastern seaboard, including some connected to Moon's Unification Church. (One estimate found that Moon's operation would lose around $100,000 a year during the closure.) Only two years after imposing limits, ICCAT delegates, meeting in 1982, agreed to once again raise the total quota of western bluefin that could be caught the following year, claiming that more tuna must be caught for "scientific reasons" so they could adequately assess the health of the western Atlantic stock. The US share of that increased quota more than doubled that year: from 605 metric tonnes to 1,387 metric tonnes. (ICCAT, after all, was a self-regulating body in which the member nations essentially made the rules, enforced the rules, and issued the penalties without any true accountability to the global environmental community. Once bluefin faded from the public spotlight, the threshold for this supposed moratorium slowly began to creep up.)

These paired moratoriums—on bluefin tuna, as well as on striped bass—soon propelled Al toward the success he had yearned for since he was a boy. Full-time fishing, as he saw it, was reserved for either those well-heeled gentlemen like Ernest Hemingway and Zane Grey or those willing to sacrifice their bodies, their personal lives, and the promise of steady income for a hard life of commercial fishing. But chasing two well-known catch-limited fish with men and women willing to pay for the privilege of going out fish tagging with Captain Al Anderson on his well-known *Prowler*? That felt attainable.

In 1981, even as Al observed numbers of "Atlantic bluefin tuna reach[ing] an all-time low," he and his clients caught, tagged, and released 70 individual juvenile tuna, a convenient way to extend the fishing day and deliver a happy angler back to the dock. But it was obvious to Al, as it was to Mather and everyone else involved in the fishery, that the idea that any tuna caught and sold counted as "scientific monitoring" was a crock of baloney. It "simply allows continued fishing for profit," Al wrote in his 1990 book *The Atlantic Bluefin Tuna: Yesterday, Today, and Tomorrow,* "and assists . . . only by profiling the continuing decline of this resource." In 1984, the Atlantic Tuna Club's policy was clear: "The catching and killing of bluefin tuna is inconsistent with the principles upon which the Club was founded and members are strongly urged to tag and release all tuna."

By 1988, Mather, who had long passed retirement, looked back with pride on his initial tagging work and the catch reductions and moratorium that eventually came about. "If we hadn't started tagging bluefin tuna back in the '50s, we never would have had the information to prove they were being over fished," he told a magazine reporter in Florida, where he spent his winters. "Without tagging, which brought management, I doubt you'd be able to catch even one off Cape Cod this summer." Mather's tagging work deeply shaped how bluefin were managed and protected. By the late 1980s, his mantle would be assumed by a new wave of scientists, including Molly Lutcavage and Carl Safina, whose rivalry would encapsulate decades of fights over the species yet to come.

MOLLY LUTCAVAGE GREW up in the mountains of Pennsylvania. Born in 1955, an era of TV dinners and McCarthyism, she was raised alongside her middle-class family's early-model television set. She stared into its cathode ray glow, watching her favorite programs, including *Sea Hunt*, a popular black-and-white action-adventure noir starring actor Lloyd Bridges as Mike Nelson, a handsome and swashbuckling scuba diver. Watching the show, Lutcavage was transported from her family's living room in the Poconos to the ocean's edge and beyond. She also adored French oceanographer and diver Jacques Cousteau's television specials, and by the time she headed to the University of Pennsylvania, she was obsessed with scuba diving, specifically the human body's ability to drop deep below the ocean's surface without the aid of mechanical devices. That undersea world and its creatures entranced her; she imagined how it would feel, having a whole boundless ocean to explore.

At the University of Pennsylvania, Lutcavage enrolled in a challenging premed program. Her goal was to become a medical doctor specializing in how human bodies can dive deep into the ocean and survive. The work was hard, but Lutcavage buckled down, and by 1976 the slight, long-haired student was working in the school's state-of-the-art hyperbaric center. Using a room-sized pressure chamber riddled with sensors, she, alongside classmates and supervisors, tested the bodies of commercial divers. It was akin to working in the space program, except the discoveries still to be made were below the earth's atmosphere, not above.

Her careful planning started to fall apart when, one day, her male mentor at the center looked her straight in the eye. "A woman will never be allowed on any rig," he told her. For the era, Lutcavage's aspirations were simply too high. The comment spurred her to instead focus on her original interest—studying the ocean and science—and she graduated with a double major in environmental studies and biology. If the medical community wouldn't have her, perhaps she could find one that would.

By 1987, after a master's degree stint in Virginia, Lutcavage had earned her PhD in biological oceanography in Florida, studying how sea turtles manage to stay alive during their long periods underwater (when resting, some can go for up to seven hours without surfacing). Soon after, she was hired to do postdoctoral work studying diving seabirds and leatherback turtles at the University of British Columbia's diving physiology lab. Working in the shadow of Vancouver's white-capped mountains, Lutcavage researched the oxygen-holding, myoglobin-rich blood of leatherback turtles, the only surviving members of a subset of ancient, warm-bodied yet dinosaur-like creatures. Lutcavage became so engrossed in her work that her supervisor, a famous diving physiologist named David Jones, jokingly called her a "turtle hugger," although he eventually found himself equally enamored with leatherbacks and their unique, light-gathering eyes. At the time, Jones's specialty was studying bluefin tuna and its warm, mammalian-like blood.

Exposure to Jones's work was the young researcher's first experience with big bluefin, and their shared qualities with leatherbacks captivated her—big eyes to spot prey deep underwater, warm blood to regulate their temperatures and extend their ranges, and extraordinary adaptations to life in the ocean. "It's a countercurrent heat exchanger of blood vessels that warms the stomach, so basically they soup up their digestive process, so they're an eating machine," Lutcavage explained to one interviewer about the bluefin's rete mirabile system. "They can handle the rapid turnover of the food that they eat more so than any other living tuna, and the enzymes in their stomach have the highest chemical rate of production." Bluefin then convert those digested fish into energy—enough to propel fish as young as one year old across the entire breadth of the Atlantic Ocean. She marveled at bluefins' abilities to survive in extreme environments and the physiological processes that allowed them to range so far beyond what once was believed possible.

In 1993, Lutcavage reached out to a friend at the New England Aquarium in Boston, curious about some of their leatherback turtle data. On the call, he mentioned they were looking to hire a scientist on contract, someone who could run aerial surveys of bluefin tuna along the northeastern coast. Could she do it? She didn't know anything about surveying for bluefin, she said, but she had done aerial surveys of turtles—how different could it be? Go up in a plane, spot the creatures, record the sightings, and repeat until you've gathered as much data as humanly possible. She moved to Boston to take the job, unaware that it would one day put her in the crosshairs of a global controversy surrounding bluefin tuna and on track to become a prime rival of one of America's most acclaimed ocean environmentalists.

Like Lutcavage, Carl Safina was born in 1955. He spent his earliest childhood growing up in a Brooklyn tenement, where he encountered a bird that left an indelible mark on his mind. One day, just past dusk, a mysterious bird with spots spangled across its belly and breast flew through an open window into his family's apartment and alighted on the furniture. For Safina, who was not yet old enough to read, the bird's arrival out of the darkness felt otherworldly. More than a bird, the creature seemed like an emissary, visiting from a mysterious parallel world running alongside his own. His father, who raised canaries as a hobby, pulled out an early field-guide prototype and identified the bird as an olive-backed thrush. The entranced Safina was raising homing pigeons in the family's backyard by seven years old.

Over three years of caring for his pigeons, Safina's flock grew to number more than 20 birds, with all their fluttering feathers, mates, and baby chicks. He watched, his face inches from theirs, as they cared for eggs, fed their young, and squabbled for roosts and primacy. He observed how their lives paralleled his own: they woke in the morning and ate; then the parent would leave, arrive home, feed chicks with a liquid called crop milk, and settle in for the evening. In 1965, his Italian American parents decided to decamp to Long Island, New York, forcing him to leave his flock behind in the city.

Despite being only an hour or two from where he grew up, Long Island in the late 1960s provided a whole new universe for Safina to explore, one of striped bass fishing with his father at their secret sandy-beach fishing spot in a town called Bayville. As a boy, he also fished for tuna off the south shore of Long Island with his father, a casual hobby fisherman. Then one day, the bulldozers arrived to destroy their private paradise to develop a residential complex, an experience that left him dismayed and devastated. "Loss of my beach amounted to expatriation," he later wrote. "And no one is so long-remembering as a refugee." Initially inspired by a high school bird-banding trip to Fire Island and later sitting in on a friend's interview into an undergraduate program studying environmental science, he had discovered the path to his life's purpose. By 1976, in his third year of university, he had grown his hair long and was helping his academic supervisor care for a relocated clutch of young peregrine falcons, a species that had been decimated by DDT across the United States. Four years earlier, American regulators had banned the harmful pesticide, which caused eggshell thinning in birds, including pelicans and eagles. For weeks, Safina waited nearby, logging notes and feeding and observing the chicks as they grew. He eventually helped release them from their nest into the sky. He threw himself into understanding and training falcons and hawks, even making his own falconry gear. In one photo, he stands shirtless on a hill, his black, curly hair cascading to his shoulders as he stares solemnly at the camera, holding a peregrine falcon like a prayer.

By the year Lutcavage moved to Boston, Safina, by then a slight, trim-bearded young conservation scientist, had worked for the National Audubon Society for more than a decade and had earned his PhD in ecology from Rutgers while working for the group. Safina, whom I interviewed at his nearby beachside cottage in December 2021, said he reviled the degree of ocean collapse he observed around him.

By the early 1980s, striped bass catches off Safina's childhood home of Long Island had plummeted by nearly 90 percent, with prices for a whole, ungutted fish reaching a previously inconceivable $3 per pound at the dock. During Safina's teen years, fishermen commonly observed huge, magnificent leaping schools of bluefin as plentiful as "blueberries in milk," in the words of one. As Safina saw it, commercial fishermen carried the bulk of the blame for the declines, particularly the five US Atlantic purse seiners that were catching, at that point, around one-third of the American annual bluefin quota.

Safina's personal passion was relaxing on and near the water, casting with his fishing rod into the waters around Long Island from boats and beaches, landing dark-fleshed bluefish and flat winter flounder, bluefin tuna weighing as much as 100 pounds and his beloved striped bass; he ate many of them with reverence. He even caught and sold a bluefin during that decade's bluefin's boom. Yet every year he went out, he remembered the teeming waters of his youth and mourned the dwindling species that seemed to swirl around him. By 1989, ICCAT's own research and statistics division estimated that the total number of bluefin in the western Atlantic had declined from more than a million total fish to a paltry 219,000. And it wasn't just the big fish that were disappearing: young fish, worryingly, seemed to be declining quickly as well.

In his role at Audubon, Safina buried himself in scientific journals, plowing through studies by scientists, including Frank Mather, and initiating conversations with legislators and other environmentalists across the country. Like Mather, he had also concluded that the commercial appetite for tuna, fueled by Japanese demand, was one of the reasons the fish had started disappearing. Between 1985 and 1990, Japan's tuna-fishing fleet caught between 670,000 and 781,000 tonnes of tuna—bluefin and otherwise—every year globally. Yet this was only enough to satisfy about three-quarters of domestic demand for tuna in the country.

In 1989, during an interview with a newspaper reporter on the topic of banning ocean driftnets, Safina gave himself a new title: the director of Audubon's new marine conservation program, a role focused on overfishing and the depletion of commercially targeted fish. Funded by a $40,000 grant from the marine-focused Packard Foundation, he had an early sense that bluefin tuna could be a charismatic fish the average American might be able to relate to, a flagship that could serve as a symbol for the ocean's majesty, diversity, and magnificence. At the time, many conservation groups and environmental agencies didn't consider fish in their campaigns, so Safina's goal was to change that. In one early interview with a journalist, wanting to come across as a strategic thinker, he framed his choice of tuna to represent the oceans' plight as a calculated one. In reality, his experiences with bluefin "meant a lot to me quite personally," he later clarified by email, "and I was aggrieved by its depletion and incensed by the negligent management and the greed that drove it." As Safina and I talked at his small wooden kitchen table, his three dogs circling our feet, I found myself marveling at how deeply the stories we tell shape the world—Safina's strategic narrative of how and why he chose the bluefin to herald his oceans campaign had influenced generations of fellow conservationists around the world.

In the face of record prices being offered for bluefin by Japanese dealers, even the organization in charge of making sure bluefin populations stayed healthy couldn't resist the fever of the catch. In 1989, the National Marine Fisheries Service issued more than 25,000 individual permits to American vessels that allowed them to catch giant tuna. It was over a twofold increase since 1983, when it doled out 11,000. "Many people simply do not realize they are fishing under a moratorium," Al wrote in *The Atlantic Bluefin Tuna*. Even still, there were more fishermen out chasing tuna, and the proportion of vessels to the number of fish caught—there were about six times more permit holders than giant bluefin caught that year—seemed statistically worrisome.

"When you get $4,000 for a fish, it's very difficult to convince people not to catch any fish," Mather told an Associated Press journalist reporting in 1986 on the bluefin's emerging plight, as the intense catch effort and high demand resulted in a whopping $15 per pound (about $40 in today's dollars). It was an amount, as one US Congressional Research Service report found, that had begun to rival prices of some categories of illegal drugs. "Quite a few of the people in this industry are afraid that if they don't get [bluefin] somebody else will," Massachusetts bluefin harpooner Ralph "Trigger" Watson once told Al, describing the early-season rush for bluefin.

With profits unlike any he had seen for a single fish, Al was among those tempted, and on days he didn't have clients he sometimes fished for giant bluefin alongside the Unification Church fleet and hundreds of others. One time, while out on an area called the Claw, near Massachusetts's uninhabited Nomans Land island, Al found himself trolling chains of mackerel, when suddenly a large pod of tuna boiled the surface of the water en masse. Hearing the news over the marine radio, boats from all around sped toward Al's patch for a free-for-all. In the melee, boats' metal outriggers snagged on each other and bent, one harpoon shattered a glass wheelhouse window, and fishermen made "numerous threats of murder" as boats nearly rammed one another. Amid the chaos, Al saw only two fish landed.

In the charter business, if a fisherman caught a giant, often he would waive the steep, hundreds-of-dollars price of the trip for his clients, or sometimes even share some of the profits he made on the fish. In the early 1980s, Al caught a tuna weighing more than 1,000 pounds, a fish that tipped the Snug Harbor scale and then broke it completely, so he never knew exactly how much it weighed. The summer his stepdaughter Susanne worked for him—he nicknamed her "Woozy" for her tendency, like that of her mother, toward seasickness—they caught a 500-pound fish that thrashed around and broke three of his ribs. He went down, as he said, "like a sack of potatoes," while Susanne stood over him, saying, "These fish are too big, Dad."

The first time Susanne worked for her stepdad on the boat, she helped their client, an exhausted father of a newborn baby, land a Japanese-quality bluefin on light tackle. The man's wife had encouraged him to go on the prebooked trip, but neither had anticipated a fight with a fish that would last four hours. As the half-delirious angler drew the fish toward the boat, Al started yelling a series of quick instructions at his stepdaughter that she had never heard before: "Use the straight gaff to snag the fish, but wait! Don't gaff it in the body because then the Japanese buyers won't want it! Hurry up but oh my god be careful." For an instant, Susanne froze, dumbfounded and scared to death. But Al's shouting roused her from her reverie, and she followed his instructions perfectly. They finally landed the big tuna, to the slumped relief of their client, and she could tell that Al was proud.

From then on, she wore multiple layers of old clothing and foul-weather pants to keep the fish slime off her legs. She hated it when their iced butterfish bait grew old and stinky, and at the end of the 12- to 14-hour days out on the water, she arrived home so tired she often forgot to eat dinner. That year, she was saving up to buy her own car, partially funded by tips that charter clients handed her at a trip's end. They heard how Al screamed and felt sorry for her, especially on the rare days when Al didn't catch a single fish and his disappointment and embarrassment swamped the boat like fog. But she also knew him well enough by that point that she understood his raging storms weren't directed at her; once she knew what to look for, she could see they pointed straight back at himself. But despite it all, Anderson's clients left happy with the fish they had caught, as well as with the stories of the salty captain that they could share on the docks.

When hitting the water, Al ran his new *Prowler* according to three general rules he'd cover with his clients before every trip. One: be prepared; two: if you don't know, ask; three: try to have fun. And then there was the unofficial motto he'd deliver with a grin as newbies got used to the boat's sway: "If you want to leave the dock, untie the boat." He encouraged his clients to tag at every opportunity, even converting those who initially only wanted a day out on the water and a bagful of fish fillets to bring home for supper. They also liked getting shout-outs in Al's regular magazine columns in *The Fisherman*. The tuna limits helped: some days Al's clients caught dozens of school-sized tuna, but under the new, higher ICCAT quotas, they were allowed to keep a few. Each year his trip notebooks filled and piled up, and his *Fisherman* columns ran with headlines like "Bluefin Scorecard" and "Wave of the Future: Tuna Tagging Update." In 1985, about 80 percent of all the recaptured bluefin tuna reported to Mather's Cooperative Game Fish Tagging Program were caught and tagged aboard the *Prowler*. The program sent him letter after letter, reams of paper, informing him of where his tagged fish had been re-caught, how far it had traveled, how big it had grown, and how long it had been "at large." Al thrilled at the knowledge, at the achievement, and at the growing pile of paperwork.

Every morning, Daryl still woke up with Al and meticulously recorded his previous day's catch and weather report in the day planner he used to track patterns of where fish schooled and on what gear and at what depths he caught them. In her looping script, she wrote about clients he refused to take out again, and about days when every single rod on the *Prowler* had a fish on at the same time. They'd drink coffee and Daryl would hand Al his packed lunch, then stand on the house's back deck waving a dish towel for good luck as he pulled away in his truck. He'd give her a soft beep of the horn and pull away into the darkness, heading toward the Snug Harbor Marina for the thousandth time. It was their daily ritual, one they repeated in their minds and smiled at during their long days apart. Daryl did most of everything that kept the wheels on the family's wagon: cooking and cleaning, paperwork and grocery shopping, planning trips and smoothing ruffled feathers in the community after another one of Al's profanity-laced tirades on the water or the docks. She didn't mind, though; she was happy, and the work kept her busy.

To supplement his income and attract new clients, Al developed a projector presentation that he took on the road to fishing trade shows and conferences. The presentation—called "To Catch a Tuna," which he'd later use as the title of his second book—featured his tuna exploits, including cutout colored arrows carefully arranged over a map of tuna schools off Block Island. When presenting on the bluefin's visual acuity, he jauntily rested his own aviator-shaped golden-rimmed glasses on the photo, a meta-joke that made him chuckle. He got a particular kick from his own photo series of fishermen doing a terrible job: One fisherman, who isn't strapped into his fighting chair properly, is tugged toward the side of the boat, the crease between his buttocks peeking out the top of his pants. Another grimaces as he holds a heavy bluefin atop his forearms like an awkward offering. In a more ominous slide, the luminous belly of a young bluefin appears shredded with a perfect half-circle. The raw, jagged flesh shows the missing chunk that a shark bit off the fish as it was landed.

On days that Al didn't have clients, or that he felt were too windy or dangerous to fish alone, he'd sit in a soft cloth recliner in his carpeted basement and eavesdrop on the marine radio. He spent hours listening to captains talking about where the fish were schooling and who was catching what, when, and how. He took notes on scraps of paper that filled the notebooks and folders

Al poses with a striped bass during his early years of operating fishing charters.

Courtesy of Daryl Anderson

While working as a high school biology teacher, Al once dressed up as "Mr. Cell" in a costume he constructed and decorated with his stepdaughters.

Courtesy of Daryl Anderson

Daryl and Al Anderson pose with a 307-pound bluefin tuna Daryl caught on September 1, 1980, during the first Annual Point Judith Masters Invitational Tuna Tournament.

Courtesy of the International Game Fish Association

This 1564 ink engraving by Georg Hoefnagel of Conil's ancient bluefin tuna beach seine fishery is one of the earliest known printed images of tuna processing in the region.

Digitized and provided courtesy of Ignacio Soto Medina, Cádiz Atlántica

This woodblock print by Utagawa Hiroshige III (1842–1894) depicts the harvesting of *shibi* (yellowfin tuna) using a net in Japan's Kyushu region. As with ancient bluefin fisheries in Spain, the fishery used a watchtower situated at the top of a nearby hill to determine the best time to haul the net.

Courtesy of the Funabashi Municipal West Library

Tommy Gifford (left) and Michael Lerner pose with a bluefin tuna in Wedgeport, Nova Scotia, in the mid-1930s.

Courtesy of the International Game Fish Association

Michael Lerner (standing) and Tommy Gifford pose with their
bluefin tuna haul in Wedgeport in the mid-1930s.

Courtesy of the International Game Fish Association

Filmmaker Margaret Perry poses with a film camera
sometime during the 1940s.

Courtesy of Leanna Griffith

In October 1979, Ken Fraser landed the world's largest bluefin tuna near Auld's Cove, Nova Scotia. The fish weighed 1,496 pounds and was more than four meters long.

Courtesy of the International Game Fish Association

This is one of the earliest-known photos of Frank Mather tagging bluefin tuna, taken by crew member and photographer Marty Bartlett in 1962.

Courtesy of the Penobscot Marine Museum

A young Carl Safina poses with a fish.

Courtesy of Carl Safina

Carl Safina, 30, poses with an Atlantic bluefin tuna he caught in 1985.

Courtesy of Carl Safina

Allan Hokanson (in cowboy hat) poses with Rev. Sun Myung Moon (far left)
alongside two of Moon's sons and two other Unification Church members.

Courtesy of Allan Hokanson

Al Anderson (left) and Jason Williams pose with a tagged juvenile bluefin tuna on September 27, 2004. It's impossible to know for certain, but this fish may be Amelia.

Courtesy of Jason Williams

Fisheries oceanographer Molly Lutcavage tags a juvenile bluefin tuna with a pop-up satellite tag (PSAT).

Courtesy of UMass Amherst

Molly Lutcavage holds a PSAT dart and tag.

Courtesy of UMass Amherst

Al Anderson poses with trophies awarded to him by the International Game Fish Association for his tuna tagging efforts.

Courtesy of the International Game Fish Association

One incarnation of the *Prowler* takes to the waves in 1981.

Courtesy of the International Game Fish Association

Rafael Márquez Guzmán wrangles a giant bluefin tuna during
Spain's annual almadraba.

Courtesy of Ignacio Soto Medina, Cádiz Atlántica

Frozen bluefin tuna line the floors of Tokyo's Toyosu tuna
auction building in December 2019.

Courtesy of the author

Three different bluefin tuna cuts served as part of the omakase lunch at Kappo restaurant in Madrid, Spain.

Courtesy of the author

This tag—National Marine Fisheries Service number R197521—was affixed to Amelia from September 2007 until she was killed in the summer of 2018, at which point the tag was given to Portuguese fisheries biologist Pedro Lino.

Courtesy of the author

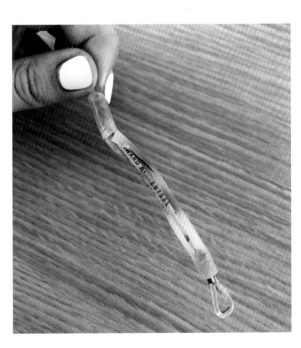

Al Anderson poses in his
home a few months before
his death.

Courtesy of Michael Cevoli

In this undated picture, Al Anderson poses with
a tagged juvenile bluefin tuna.

Courtesy of Daryl Anderson

he filed alongside stacks of scientific papers and letters to and from Mather. He never threw away envelopes. If he got mail, he'd open the letters and neatly stack the envelopes near his easy chair, so when he had a new idea, he could scribble it down. The resulting scraps drifted all over the house: new book ideas, plans for a new tagging-flag design, something to do on the boat. In fishing season he'd go to bed early; on other nights he'd watch science documentaries and old movies. Most nights he'd have a few Jim Beams in the evening to relax.

In the wintertime, to stave off boredom, Al refurbished his rods and reels in his basement workshop, like he had learned from his uncle Harry, wrapping the fiberglass rods carefully with tape and thread, oiling the bearings and greasing the gears. His step-daughters sometimes joined him here, watching him work and asking questions, sometimes just listening to him ramble on about fish, where they went, what he'd do better or differently next season. He designed and bolted a wooden reel the size of a basket-ball to his workbench that he could turn with a crank to add and remove monofilament lines from the rods. As snow blanketed Snug Harbor, Al would rebuild every single reel he owned, lining up boxes of metal reels, cleaning them with a cloth, and going over their metalwork. Each one was carefully wrapped so he'd know he hadn't missed any.

Most fishermen paid shops and other gear experts to respool fishing line, as the slightest fray or wear in the braided high-gauge monofilament—a space-age material thinner than angel hair pasta that can tow a 1,000-pound fish—meant losing a $10,000 fish. Instead, Al bought line by the boxful and spooled it himself on his wooden reel. Standing in the warm garage, Chuck

Berry and the Everly Brothers and Patsy Cline piping from the black plastic boombox perched on a window ledge, he turned his hand-built cylinder by its small wooden handle: 12 turns for 25 feet of line, 96 turns for 200. Sometimes he enlisted his step-daughters to turn the reel. One of his tricks was to use knots he taught himself—knots it seemed only he knew—to tie markers onto the line: green, orange, and red, so when he or a client was fighting a tuna, he would know exactly how much line had to be reclaimed back onto the reel. Red meant the tuna was far away; green meant the fisherman was in the homestretch. Some fisher-men who knew him were amazed at the quality and reliability of Al's work on his rods. "Never mind that I couldn't do it," said Al's former fishing mate Matty DiMatteo. "I couldn't even find someone else who could do it." Al's garage and workshop kept fill-ing up with rods and lures, bits of wire and scraps of plastic. Some days the phone rang off the hook with bluefin sportsmen looking for a bite.

Every day on the water he'd land his shiny new *Prowler* at his regular berth, near the gas pump at the gray-planked Snug Har-bor Marina, then reveal with a flourish the fish his clients were legally allowed to keep, cleaning them right there on the dock. He'd skillfully separate fillets from the bones, chuckling inwardly as clients from other boats gawped at his catch and resentful cap-tains glowered at how paltry their catches often looked in com-parison. Every so often he'd catch a glimpse of a tag hanging from a tuna suspended at the docks, unspotted by the triumphant angler. Whenever he noticed, Al would point out the tag and sug-gest the angler report and return it to the issuing agency, a simple matter of mailing the tag along with some details on where the fish had been caught and how long and heavy it had been. Many heeded the advice, although one time an angler was bold enough to voice a sentiment that many others held: he simply "couldn't be bothered."

IN THE FALL of 1989, Daryl got a call from Al, who was on the road for one of his lectures on catching and tagging bluefin down in Long Island. Somehow, during the drive north, he had blown past Narragansett by an hour and a half and ended up outside Boston. "I don't know how I got here," he told Daryl on the pay phone, his voice strangely vulnerable. "Well, do you know how to get home?" she asked him. "Yes," he said, pausing, his characteristic wit starting to stir, "but it's going to take a while."

A few weeks later, Daryl got the report from their family doctor and asked him if she could break the news to Al herself. He had a brain tumor. "What are you talking about?" he asked her, shaking his head in disbelief. Spring would be coming soon, and he had to run the *Prowler.*

Together, they went in to meet the specialist who would be doing Al's lifesaving surgery. The tumor was in the center of his brain on the end of a long stalk. Every time the blood moved in a certain way, Al would lose bursts of short-term memories, and surgery was the only option. Sitting in the doctor's office, the couple watched carefully as the specialist pulled out and took apart a model of a human brain and explained what would have to be done. As the doctor explained the recovery process, he struggled to fit the plastic brain pieces back together. Giving up, he tossed the pieces behind him on a shelf. Al looked at Daryl in horror. As they climbed into their car, Al finally exploded. "Did you see that?" he ranted. "He doesn't know what he's doing!"

The week of the surgery dawned and, with no opportunity to pick a new doctor, Al went ahead with it. Daryl waited, outwardly stoic but her heart on fire with worry. Finally they called her in: the surgery had been a total success. Al would need a mate for the rest of his fishing career, and he'd need to stop drinking due to the anti-seizure meds, but he was going to be fine. And when he came home, to Daryl's relief, he was so much easier to be around.

Things were good, they were calm, and a decade of Al's most productive tagging years were right around the corner. It had been a health scare, that was all, Al told himself. And the world had only seen the start of how many fish he could tag.

CHAPTER EIGHT

THE EXTINCTION AGENDA

~~~~~~~~~~~~~~~~~~~~~~~~~~~~

Al, 1990s

Man is inherently a restless remaker of his own world.

—PHILIP WAGNER, *THE HUMAN USE OF THE EARTH*

Sitting in his carpeted basement while recovering from brain surgery, flipping through letters exchanged with Frank Mather and catching up on the latest fishing news, Al carefully followed the troubling collapse of bluefin stocks in southern Europe and northern Africa with concern. Despite the line drawn down the Atlantic by ICCAT a decade earlier, his own tags had already shown that fish that traversed American waters could end up on the other side of the ocean, where they were being killed and poached with increasing frequency. But despite that— or, at times he wondered, because of it?—the charter business he and Daryl had worked so hard to establish was thriving, and catching and tagging the quota-capped, hard-fighting bluefin was increasingly why clients were searching out the *Prowler.*

As Al recuperated through the winter of 1989, Daryl cleaned the house, cooked the food, washed the clothes, continued working at the marina, and maintained correspondence with clients who were excited to fish with Al in the upcoming season. Al had healed quickly enough for his doctor to reluctantly concede that he could take to the water again in July 1989, although Al continued to grumble about his doctor's uncompromising order: he wasn't allowed to fish without a first mate aboard. Mates could be helpful, but they also could be a giant pain in his ass.

Still, now with another angler aboard, typically someone like the young Steve Tombs, who had worshipped the skipper and proved capable of absorbing his angry tirades, Al would head out on the *Prowler* seven days a week when conditions were right and fish were biting. Sure, there seemed to be fewer bluefin out on the water, and the ones they were catching seemed to be getting gradually smaller and were harder to catch reliably, but Al was good enough that his clients rarely headed home without a good story and at least a fish or two for dinner. He also had found increasing satisfaction from writing books about fishing, the first of which was picked up by a regional publisher interested in capitalizing on the bluefin boom.

Book research took Al to places he'd never expect to end up. One example was his meeting with Brigs Endt, the captain of a longlining vessel named the *Catherine E.* Al met Endt one March when the younger captain found himself in Rhode Island with a boat that needed some minor repairs during the fishing off-season. The pair got together at George's of Galilee overlooking the Point Judith Harbor of Refuge and shared a seafood dinner while Anderson peppered Endt with questions about longlining that he hoped to print in his book. During their meeting, Endt told Al a troubling story—not about a fish that got away, but instead about a group of bluefin that couldn't escape:

Around Thanksgiving 1988, the *Catherine E* cast off port from Montauk, New York. Endt piloted his vessel for the Hudson Canyon, a two-mile-deep submarine basin 160 kilometers southeast of New York City, hoping for a good catch of yellowfin or bigeye tuna. Both fish earned good money at the docks, with bigeye fetching as much as $18 a pound, or a handsome $45 in today's dollars. Endt needed a good paycheck to offset the costs of running his expensive longlining business. Between materials and equipment, it wasn't unusual for him to spend $10,000 preparing for a single 18-day fishing trip, and that didn't include any labor.

That crisp fall morning, the *Catherine E*'s gear had been set all night. The main line, which could be up to 55 meters long, was suspended at the surface by bobbing buoys. And along the line, hundreds of baited hooks floated at the depths where pelagic predators like swordfish and bluefin hunted. Endt and his crew had released their gear the evening before, baiting its hooks with the biggest squid they could buy, each about 20 to 22 centimeters long. Just before sundown, they laid out the gear and let it drift, and then followed the signal of beepers attached to the gear so they could find it and then haul it in, which took six or seven hours.

When Endt's crew pulled in their lines that morning, they were dismayed to discover 28 bluefin hanging off their third section of gear. Endt knew the US longlining fleet had already reached its total allowed catch of bluefin tuna a few months earlier, back in July, which meant it wasn't legal to keep the valuable fish. But by that point, most of the fish were likely already dead after being dragged through the ocean for hours, fighting and failing to free themselves.

As the longlines were retracted onto their enormous metal reels, each fish of the catch breached the surface headfirst, drawn upward toward the ship's deck by the screaming line. Before it could be landed, however, each fish thrashed its tail vainly in the air, eventually slamming into the boat's broad metal side and upward toward the crew that waited to unhook it. Hit after hit, the volume and tone of their bodies striking the boat's flank varied by species and size. It was a grab bag of albacore, bigeye, and swordfish, creatures that would be frozen and sold when the boat eventually arrived back at shore. But of those dozens of bluefin, those gorgeously striated fish with their chartreuse finlets and gasping, bullet-like heads, not a single fish could be kept. The crew, cursing the time they had wasted on those giant fish, cut them away. A few fish, Endt said, might have been in good enough shape to survive, but he'd never know once they disappeared under the surface. (Recent research in the Gulf of Mexico suggests that all those fish, had they survived, likely would have died of heart attacks after having been dragged in warm surface waters for an extended period of time.)

"No way I was going to have any fish in my possession after the closure of the fisheries," Endt lamented to Al as he remembered the pointlessness of catching those 28 bluefin. "Fellows on deck knew it too."

IN THE YEARS following Al's recovery from his brain tumor, life had outwardly returned to normal for the Anderson family. But early one evening in 1995, shouts tore the air of Al and Daryl's quiet community, exploding a secret the couple had tried to keep from their tight-knit social circle for more than a decade. Since he had retired from teaching, Al had struggled with a gradually intensifying drinking problem. Only Daryl and the girls knew how he could get off-season during the long, cold Narragansett winters, bored and isolated, drinking Jim Beam downstairs by himself. To Daryl's great relief his binges had stopped, back when Al wasn't able to drink because of his brain tumor medications. But on their first trip down to Florida for a vacation a few months after his invasive surgery, Al had headed straight to the package store to buy bourbon. Daryl couldn't believe it. There they were, having made it through Al's brain tumor scare, but Al was drinking again, this time in a way that tested his otherwise indefatigable wife. And when Al drank alone, he often did so with a frequency and force that worried her.

On the evening that everything fell apart, Al came home, walked into the basement, and had one drink, then another, and then another and another. Daryl realized that she had to put her foot down and make it clear to Al she wouldn't put up with it anymore. It had just been too much of the same thing for far too long, she thought. She confronted him in their kitchen, telling him he had to stop. To her horror and surprise, he exploded, transforming into a mass of tall, shouting anger and fury, the kind he would show on the *Prowler* only on his very worst days.

Scared and panicked, Daryl fled her own home, the house her parents had bought for her after her first marriage disintegrated, and where she and Al had built their lives together. She ran a short, jolting distance to her neighbor's doorway. Standing inside that unfamiliar home, out of breath, Daryl said the words out loud, hearing herself as if removed from her own body: "He's out of control," she sobbed. "I can't control him." The neighbor, who worked as a parole officer, picked up the phone and called 911.

Within minutes of responding, the police arrested Al, dragging the big, protesting fisherman to the local precinct, as Al relentlessly insisted it had all been a terrible mistake. He spent the night in a cell, a drunken, rumpled, embarrassed tumble. When his lawyer picked him up the next morning, he informed Al that he couldn't go home. Daryl packed Al's possessions into boxes and his clothes into a suitcase, and the then 56-year-old fisherman was forced to move back into his mother's nearby house. Al's outwardly orderly life—one where he had savings in the bank, a boat to fish from, people who loved him—had, once again, slipped away. He lay in bed in his mother's home, staring at the ceiling, his thoughts running wild: it was all his fault; he would never escape the terrible darkness.

After his case worked its way through the local legal system, as mandated by the court, Al attended a handful of counseling sessions and even mustered the courage, on the advice of his counselor, to attend a few Alcoholics Anonymous meetings. He loathed those badly lit gatherings and what he saw as their boring, sanctimonious guidelines, the pathetic self-pity and the excuses. Daryl went to Al-Anon meetings of her own, sharing her story with other men and women whose spouses, parents, and siblings had been pulled away from them by forces they couldn't quite fathom.

After that night in jail, Al made a decision, the only one he could live with: he would simply stop drinking. There were no momentary lapses after that, no "I'll just have one and then stop." He decided overnight and never had a drink again. After a month of living apart, Al asked his lawyer to pass along a message to Daryl. He had stopped drinking, he told her, and he wanted to return home. She was reluctant and desperately unsure, but she missed their partnership. So she agreed to give him a second chance.

At Daryl and Al's first marriage counseling appointment, in sessions led by an older couple within their church community, it was obvious Al couldn't stand being there, and those sessions didn't take. So Daryl and Al figured it out on their own. They talked, they negotiated, Al repented and apologized, and they figured out what they had to do to rebuild their lives and Daryl's broken trust. And Al proved true to his word: he not only never had a drink again, but for good measure he also stopped smoking. The couple's neighbor, the probation officer who had called the cops on Al, was gobsmacked by the sudden change. "I've never heard of anyone doing cold-turkey smoking and drinking," she told Daryl. "Most of the people I know, they relapse."

Once Al had settled back into the house, a new energy imbued the carpeted basement, where he still held court with fishing friends and wove tales for his grandchildren, where he watched nature documentaries and conducted his correspondence. "The shift was abrupt and total," fellow fisherman and former mate Zach Harvey observed of Al's shift from fishing for food to fishing for conservation and science. Al had been given yet another precious chance and this time he was going to make it count for something. His charter business was growing, and return customers continued to call every season, excited to get back out onto the water for another exhilarating day of catching and tagging fish. Enough clients were interested in the tagging aspect that Al decided this was how he would make his comeback: he was going to make his fish tagging the centerpiece of his fishing career, evangelizing for the practice while encouraging other fishermen to help draw attention to a species hurt by rapacious fishing practices and a climate that had obviously started to warm in front of their eyes. Tagging might hold the key to the bluefin's survival, so Al would encourage his acolytes to continue releasing the fish back into the ocean.

WHEN MOLLY LUTCAVAGE first started working with bluefin in the early 1990s, many fisheries scientists believed that the breeding population of the species in the western Atlantic was in the mere tens of thousands, a 90 percent drop from 1975 levels. (Although some scientists felt that early catch data indicated the population's peak had once been many magnitudes higher than that before the Japanese longliners arrived in the 1970s.)

In the early 1990s, the East Coast Tuna Association, which financially supported some of Lutcavage's research and had paid scientists to consult on the association's behalf since the start of the moratorium, started championing the idea that a natural phenomenon called "regime shift" was to blame for the dropping bluefin numbers. It wasn't that tuna numbers were declining, the ECTA argued, but that the same fish were heading to northern Europe, migrating and spending time near Norway and Sweden instead. This argument later evolved into another contention that due to overfishing of bluefin tuna on the European side, fewer young eastern Atlantic bluefin were making their transatlantic voyages to the west. It wasn't that American fishermen were catching too many fish, the ECTA argued, but that overfishing elsewhere skewed numbers on "their" side of the ocean. Rich Ruais, who worked for the ECTA at the time, said he noticed that most fisheries scientists "didn't know jack" about the bluefin tuna's life cycle, which meant most conclusions and fisheries management decisions they were making were, at best, calculated guesses. For a multimillion-dollar American industry, he thought, that simply wasn't good enough.

It was this precipitous drop in his local bluefin populations, as well as his own love of catching bluefin from his Montauk vessel, that had compelled Carl Safina to rally on behalf of the species. Within a few years, his advocacy caught the attention of influential bureaucrats within the American fisheries system. In 1991, the US secretary of commerce appointed Safina to the Mid-Atlantic Fishery Management Council, an influential regional position he eventually held for three years. Also in 1991, as part of his Audubon campaigning, Safina drafted a formal petition to the treaty members of the Convention on International Trade in Endangered Species of Wild Fauna and Flora (CITES), proposing that the bluefin be listed as endangered and that a subsequent ban of international trade in the species be instituted. He wrote the petition after both US and Japanese ICCAT representatives told him, during one international meeting on the fish, that despite the dire predictions of their own scientists, they had absolutely no intention of reducing their own bluefin fishing quotas.

If it were listed under the treaty's most restrictive level,

Appendix 1, the doors of the booming global trade in tuna would slam closed. Without those thousands of giant fish flown every day from the Canadian Maritimes and US Northeast to the Japanese market, the fishery would be hamstrung. Safina hoped that would give bluefin populations time to recover. His proposal was the first time someone had suggested protecting a commercially exploited fish under the treaty, and it sparked an inferno in the bluefin scientific community.

Lutcavage, meanwhile, was trying to find and record data that, to her mind, could push back against the unscientific and radical environmentalist agenda championed by Safina and others. It was one thing for policymakers and ideologues to debate the fate of the species, but the amount known about bluefin was still frustratingly sparse, given its value and importance in ocean ecology. Working alongside Scott Kraus, the marine mammal specialist who brought her onto his team at the New England Aquarium, she had pursued first-of-its-kind research gathering and analyzing data on bluefin. When she started her contract, federal fisheries managers weren't yet running their own aerial surveys, but the commercial fishermen were, paying spotter pilots—whom Lutcavage considered "brave guys" with fuel tanks strapped to tiny aircraft—to find and direct their vessels to bluefin schools from the air. Spotter pilots at the time were considered part of a boat's crew and typically received a 25 to 30 percent cut. The ECTA told Lutcavage that government fisheries weren't willing to fly with the spotter pilots to see the bluefin abundance with their own eyes. So how, she wondered, could the regulators possibly know that the bluefin population was in trouble? As she watched the bitter feud between fishermen and environmentalists like Safina escalate, she couldn't help questioning the latter's motives, which she dubbed their "extinction agenda." If their careers hinged on fighting ongoing threats to specific species, when those threats disappeared, they'd lose funding or even their jobs, so they were, she felt, under pressure to keep the bad news coming. From her vantage point, while bluefin were facing challenges of overfishing and criminal activity, her time spent talking with fishermen on the docks made it hard to take Safina's accusations seriously.

From 1993 onward, Lutcavage spent fishing seasons flying with spotter pilots and joining purse seiners on their harvesting trips, research that was hard but fulfilling as she collaborated with fishermen she respected and whose company she enjoyed. When not at sea, she worked at her office, a glorified closet at the aquarium, where she was eventually joined by an assistant. Together, they projected slides of tuna schools against the wall and painstakingly counted by hand the numbers of tuna in each group. "We had upwards of five thousand surface individuals," Lutcavage later recalled of the groups. Using available estimates, they were told by the spotter pilots that this meant there were likely thousands more large bluefin below the ocean's surface. "We knew exactly where the schools were photographed, because we had the LO-RAN [long-range navigation] and GPS," she said. Lutcavage gave her finished reports to the fishermen who had allowed her aboard their boats and planes, and they, in turn, used her findings in meetings with congressional staffers and fisheries managers to argue for higher quotas. Frank Mather had once posited that, because humans can count only the fish they catch, the youngest bluefin of every spawning year would inevitably be the species' "cryptic biomass"—meaning scientists' data on the growing health of the species would always be a year behind. Lutcavage thought this theory had merit. What she was witnessing through her work, Lutcavage believed, was proof that the doomsday scenarios championed by Safina and his ilk overstated the problem. How could they possibly know how bad things were if they weren't doing research themselves?

Lutcavage continued earning the trust of the notoriously truculent fishermen of Gloucester's historic docks, which included running data-gathering operations aboard massive purse seiners. To her mind, she was following in a tradition of fisheries scientists embedded within industry. Working on their own, government scientists didn't have the money to pay for those far-ranging cruises and the thousands of dollars of diesel and the enormous ships required to travel so far offshore in their pursuit of tuna schools.

The commercial fishermen and their large boats had increasingly developed their own ingenious ways to follow ICCAT's rules. Some Atlantic purse seining captains painted a two-meter-long life-size red silhouette of a giant bluefin on their vessel's deck so spotter pilots above could use the outline to determine the true sizes of the fish they were seeing. This "ground truth" meant the captains could target giant tuna that length or longer or, alternatively, schools made up of smaller fish that were allowed to make up to 10 percent of their catch. Now savvy navigators of the hot bluefin market, those captains had learned that if they caught too many fish or too many giants, they could risk flooding the market and tank the prices they could earn for bluefin.

The astronomical value of bluefin also meant that both Canadian and US fishermen increasingly found themselves at odds with both fisheries managers and scientists, which for many was a new and unpleasant situation. Throughout the 1960s and 1970s, they had largely been on the same side, when their priority had been keeping foreign fishing fleets out of North American waters. But ever since the ICCAT-imposed scientific quotas of the early 1980s, Lutcavage had noticed that distrust between the two groups had been growing. From the sidelines, she listened to reports of how her data—which she believed demonstrated that the Atlantic bluefin population was far stronger than environmentalists like Safina claimed—was being ignored in lieu of cutting fishing quotas. It seemed to Lutcavage that the science was being undermined by environmentalists' ungrounded and unscientific accusations that disrespected her work. It also bugged her that her scientific credibility was being challenged simply because she was gathering data alongside fishermen who lived and worked in communities where bluefin income supported whole swaths of the economy. Of course, she said, they had a vested interest in the survival of the species. "I was outraged by the lack of interest in getting the science right," she said. "The whole idea of cooperative research partnerships with fishermen is what the old guys did. There was nothing wrong with that." There were problems in the bluefin industry, to be sure, including overfishing, illegal harvests from boats off the Faroe Islands, and unregulated fishing across remote Mediterranean ports and North African countries, but she believed there was no way bluefin should be listed as endangered. Yet that was exactly what environmentalists at the time insisted was urgently needed to prevent the extinction of the emblematic fish.

In 1993, the year Lutcavage started working with the New England Aquarium, she and Safina collided. By then, the ECTA, which had been founded in response to the bluefin moratorium a decade earlier, was funding the New England Aquarium's aerial surveys of bluefin tuna. The association had recently hired former New England Fishery Management Council employee Rich Ruais as an advocate, and he had poured the group's resources into independent science like the bluefin-counting project Lutcavage had been hired to oversee. Doing research that was among the first of its kind in bluefin, Lutcavage reapplied her background in tagging and tracking sea turtles to try to better understand the giant fish, and collaborated with state scientists and tuna association harpooners to jam sonic tags into 12 individual fish. The tags sent electronic pings, like a bat's echolocating chirps, that allowed Lutcavage and her team to mark and record the location of each fish that carried a tag. They were designed to track the lightning-fast movements of the bluefin schools but had only a one-kilometer range, which meant Lutcavage and her collaborators had to stay awake for 48 hours at a time, chasing the tuna from the surface as they scrambled to keep the fish within range. It was exhausting, exacting work, but Lutcavage was exhilarated by the opportunity to glimpse things no one had ever seen before, including how bluefin moved and traveled underwater, where they were going, and how they congregated along the US coastline.

Safina caught wind of the work and, according to Lutcavage, sent a strongly worded letter to New England Aquarium conservation director Greg Stone. (Safina has no recollection of sending the letter, and maintains the first time he met Lutcavage in person was at a scientific meeting where she refused to look him in the eyes.) Lutcavage perceived Safina's letter as a clumsy attempt to discredit her lab's aerial surveys, and as evidence that her group couldn't "work in research partnerships with fishery stakeholders without being duped, or used," she observed.

And so the stage was set for a long-standing conflict between the two. Safina, compact like a jockey since his days growing up in Brooklyn, realized he would always lose in a fistfight so resolved to win fights with his mind. Tipping himself into the fracas of working to protect bluefin, as he told me at his Montauk cottage, he intentionally assumed the persona of a "forceful advocate" for the fish—one who would never, ever be steamrolled. For years, he hammered everyone around him with the same two simple ideas: that fish counted as wildlife and that humanity was witnessing the depletion of the bluefin just as it had once presided over the collapse of the Great Plains buffalo. He channeled his love of nature into being its protector and champion, attending dozens of meetings armed with a briefcase full of scientific papers and memos. There he scrapped with his opponents—people who were, in his view, bullies or liars and sometimes both.

In every meeting, one of the most grating voices in the room, he felt, belonged to Rich Ruais. For his part, Ruais thought Safina was an opportunistic grandstander, a grasping leftie who had hitched his wagon to bluefin tuna and decided to ride the money train as far as it would lead. Ruais would often attend meetings in New York and New Jersey just because he knew Safina would be there. At meetings, the two skirted around each other, never shouting but always adding a pointed barb here or there. In his first book, Safina described Ruais as "a tense man in his early forties with a meet-the-Beatles haircut and a thin, flat mouth." One day, after a random check by authorities caught Safina with an undersized striped bass in his boat—which Safina says was caught and mismeasured by a couple of centimeters by a friend that was later reported by an industry publication—Ruais made a point to bring it up at every opportunity. They were like oil and water, both convinced they occupied the moral high ground, both convinced the other didn't know what he was talking about.

As consultation meetings on bluefin slowly spread across the Northeast, coordinated and hosted by the National Marine Fisheries Service and held in community halls and other public venues, Safina pulled up a chair at as many as he could. He talked about the tables of data gathered by scientists that showed the steady downward march of how many fish were being caught, and about how few fish were likely left. These findings directly contradicted what Molly Lutcavage believed she was seeing out above the waters of New England. It was her studies that Ruais waved in the air. Everything was fine, he insisted. Everything wasn't fine, Safina argued back.

Safina diligently attended these meetings, one after the other, making his case for the huge fish. From his perspective, few attendees had bluefin's best interests at heart. The fishermen called themselves "stewards of the resource," but most days Safina had the clear sense that they were primarily stewards of their own bank accounts and kids' college funds. They were willing to protect tuna in only the most minimal ways, and viewed the problem through the lens of their own short human life spans. Safina was there to fight for the future. But the meetings wore on him. As soon as he'd get back to his hotel room after a day of debates and paperwork and petitions, he would take two aspirin and slip into a hot bath, trying to soak away the stress.

Once, when Safina planned to take his new seven-meter Grady-White fishing boat out in Montauk, a fisherman walked toward him on the dock. "Are you Carl Safina?" the burly man demanded roughly. "I am," he replied, extending his hand tentatively for a handshake. Instead of taking his hand, the fisherman exploded into an angry rant, hurling profanities at Safina as wide-eyed observers watched the scene unfold. "If I ever see you around this dock again . . . ," the glowering man threatened, seemingly unaware that Safina had frequented and fished from the marina for decades. Instead of heading straight out to his boat and onto the water, Safina hung around the dock quietly until the fisherman grumblingly drifted away. He didn't want the man knowing which boat he owned: it would be easy for him to return in the night and drill a hole into its hull, sending his new boat to the bottom of Block Island Sound.

The violence and vitriol exhausted Safina, who felt the tension creeping into his own mind and body. Some days he wondered if the endless, stressful meetings were changing any minds, when it seemed obvious that most people out on the water knew there simply weren't as many bluefin as there used to be. "There is no comparison between then and now," one fisherman told a *New York Times* journalist in 1991. "The best guys around here are lucky to get eight or nine giants a year. We used to get seven or eight a day." Safina, interviewed in the same story, recalled listening in on the VHF marine radio as one skipper chided another fisherman, encouraging him to leave some bluefin in the water for another day's catch. "Nobody left any buffalo for me," the disembodied voice on the radio quipped, a comment that inspired the analogy Safina used from that day forward.

At one particularly heated encounter at an early-1990s NOAA-hosted meeting that Safina attended, two bluefin fishermen of different fishing-gear types got into each other's faces. They threw punches at the front of the room, and instead of kicking the pair out, the gathered group eventually broke up the fight. As I imagine that fight, perhaps it felt cathartic and good for some in the crowd, those onlooking men who watched their pain, frustration, and fear displayed in a language so visceral and concrete. For Safina, it felt like a sign that his presence on the committee was theater and a waste of his time.

Despite getting initial positive feedback from the US delegation to CITES for Safina's petition, American fisheries managers eventually declined to support the proposal to list the bluefin as endangered. For months, commercial fisheries groups, including the ECTA, paid lobbyists and scientists hundreds of thousands of dollars to support their case to the states and the federal government, and their efforts were rewarded. Despite the setback, Safina didn't give up. On the advice of Swedish World Wildlife Fund (WWF Sweden) employee and Swedish fisheries board member Lennart Nyman, Safina and his colleagues instead convinced the Swedish delegation to back the proposal at the 1992 annual CITES meeting in Kyoto, Japan.

At the closed-door meetings, the Japanese diplomatic delegation—made up of 49 members, compared with most countries' handful of delegates—and Canadian and Japanese political lobbyists worked frantically to scuttle the bluefin listing. At the conference, Safina was furious to see a Japanese- and ECTA-sponsored information booklet sitting at the official ICCAT information booth alongside graphs that showed a line steadily progressing upward. Safina found it a Kafkaesque twist that instead of showing populations going up, which he believed was the desired impression, the graphs instead visually charted the banal fact that as bluefin aged, they *grew larger.*

Instead of moving ahead with its proposal, Sweden eventually offered to back down if ICCAT member countries including Japan and Spain agreed to sweeping bluefin catch reductions, which they publicly did. So Sweden withdrew its proposal, and the chair squashed all further debate on the topic. "There is no record of what the Japanese did to the Swedes to bring about this withdrawal, whether it was threatening not to sell them Toyotas or Sony TVs, or perhaps erecting more tariff barriers to Volvos," wrote Charles Clover in *The End of the Line.* "But it worked."

In the aftermath, according to environmental governance professor and *Red Gold* author Jennifer E. Telesca, a furious Carl Safina alleged that the Spanish delegation had threatened Sweden with exclusion from the fledgling European Community unless it backed down from its proposal. Before leaving the meeting, an American lobbyist working for the Japanese fishing industry in Washington, DC, approached Safina and gave him a friendly squeeze on the arm. "I want to thank all the conservation groups for making this a lucrative year," Safina said she told him with a wink.

Notwithstanding the petition's challenges, the pressure and lobbying by Safina and others did accomplish something: in November 1991, with the threat of a CITES listing still hanging over their heads, ICCAT member states had agreed to a 25 percent quota reduction in the western Atlantic over four years. Safina, who didn't trust promises made by Japan and Canada in

Kyoto, accused ICCAT of "negligent management" in a June 1993 article in *Conservation Biology*. In the article, he dubbed ICCAT the "International Commission to Catch All Tuna," which was eventually tweaked to "International Conspiracy to Catch All Tunas," which caught on among environmentalists. As far as they were concerned, any mandated reduction in catches by ICCAT counted as a win, and was a sign that their environmental campaigns at home and abroad had begun to bear fruit. In 1992, spurred on by increasingly dire population predictions for the species, ICCAT agreed to a 50 percent reduction in catch quotas phased in over two years.

That significant reduction infuriated commercial groups, which, in their multipage responses to regulators and politicians, invariably managed to include a scathing section not so subtly directed at Safina and his bluefin-hugging cohort. "As far as we can see there is not a competent and knowledgeable bluefin scientist in the world that would lend their name and good credentials to the suggestion that bluefin tuna in the Atlantic Ocean (East, West or Mediterranean) is threatened or about to be threatened with extinction in the foreseeable future," wrote ECTA president Stephen Weiner in an April 1992 letter to the US Fish and Wildlife Service head John Turner in the wake of the narrowly avoided CITES listing. "It no longer matters whether the [National Marine Fisheries Services]'s motivations for destroying our traditional commercial fisheries are fueled by a desire to provide higher allocations and availability to a recreational constituency or a sincere desire to accomplish the fastest possible recoveries," he wrote. "The bottom line is that these policies are inconsistent with President George Bush's stated environmental policies that call for a reasonable balance between environmental concerns and the need to preserve jobs for Americans."

Around 1998, Safina stepped down from the US bluefin advisory committee, deciding that his energies would be better spent fighting for bluefin in a forum where his pleas could have a wider impact. The year before, Safina had published his first book, a nearly 500-page epic entitled *Song for the Blue Ocean*. It introduced bluefin to the world in a way the fish had never been seen before and gave the Audubon ecologist a forum to recount his tribulations and disappointments championing bluefin on the international stage. On its first page he described one close encounter he had while he was out fishing as a giant bluefin leapt out of the ocean beside his boat. "[It] hung suspended for a long, riveting moment, emblazoned and backlit like a saber-finned warrior from another world," he wrote, "until its six hundred pounds of muscle crashed into the ocean like a boulder falling from the sky." *Song for the Blue Ocean* documents Safina's travels across the world studying bluefin but also Pacific salmon in the American Northwest and endangered reef fish in Palau and the Philippines. How dare we wipe them out, those magnificent creatures we had yet to understand, alongside so many others? he asks his readers. What right had we to do so?

"We have yet to extend our sense of community below the high-tide line," he wrote, calling for the formation of a "sea ethic" like that of the land ethic that birthed the US conservation movement in the late 20th century. "Simply by offering the sea's creatures membership in our own extended family of life we can broaden ourselves without simplifying or patronizing them." Upon publication, a *New York Times* reviewer declared *Song for the Blue Ocean* a "landmark book"—and it went on to be a bestseller.

IN 1996, MOLLY Lutcavage, still searching for better science and reliable data on bluefin, reached out to a physicist-biologist named Paul Howey. Five years earlier, Howey had released an animal tracking device called a Platform Transmitter Terminal, weighing less than 100 grams. The miniaturization of high-tech sensors and processing systems using pieces of plastic and soldered metals had transformed banking, industrialization, and communication across the planet, and wildlife science was the next field ripe for modernization. Prior to Howey's tags, only large birds like bald eagles and trumpeter swans—creatures large enough to bear the weight of heavier portable computers that were able to communicate with satellites—could be tagged. But his lightweight invention opened whole new fields of study for avian researchers, including those studying the endangered bald eagle. So, he reasoned, why not design something similar for fish?

This inspired the development of a device Howey called a "pop-up satellite archival tag" or PSAT, which could track fish and sea turtle species that don't surface often enough to provide reliable location data. While it wasn't the first time satellite tags had been deployed in the ocean—that work had started in the late 1970s—it was a huge leap in the science of tagging, which had largely been data-limited to information drawn from a single moment when a fish was caught, recorded, and released, unless, like Amelia, the animal was later recaptured. Howey's tag, which had enough battery life and storage for about two months, was connected to an implanted dart at the base of the fish's second dorsal fin. The device itself trailed behind the fish, taking the temperature of the seawater, which it stored in its tiny computer brain. Once it hit the time at which it was programmed to release, the device put out a burst of electric current that, due to the conductivity of salt water, corroded and broke a small piece of stainless steel wire, sending the tag bobbing to the ocean's surface. Here, the tiny device—now a literal needle in a big, watery haystack—would then upload the compressed data to a waiting satellite. It was better if a researcher could retrieve the physical tag, but the compressed data was often more than enough. Upon receiving her first report compiled by Howey and his team, Lutcavage became one of the world's first researchers to digitally track bluefin in the open ocean.

But another bluefin tuna researcher had already discovered the potential of Howey's tags. About six months before Lutcavage started using Howey's electronic tags in bluefin, a scientist named Barbara Block got to his technology first. Block was a protégé of Woods Hole Oceanographic Institution scientist Frank Carey, who had first identified how a tuna warmed its blood and later used transmitters inserted into hooked tuna to prove the fish could regulate its body temperature. In 1973 his team also proved that a bluefin could increase the temperature—and thus, performance—of its brain and eyes. Like Carey, Block had big dreams of how far bluefin tuna could take her.

~

Like Lutcavage, Block was a marine physiologist who was particularly fascinated by how bluefin lived and moved in the open sea. After earning her PhD in zoology at Duke University, Block started working at Stanford University's lab on California's Monterey Bay, where her specialty became "fish and chips," as she once wryly observed during a presentation at an American Association for the Advancement of Science's annual meeting. That is, she had carved a niche for herself as the world's leading expert in tracking huge bluefin across the open ocean, using new microchip-based technologies. According to Howey, Block first heard about his work and reached out to him, excited to deploy the devices, which he priced at about $3,600 US per unit. Within polycarbonate casings, Howey embedded tiny microchips that communicated with satellites via thin antennae, and they provided a perspective on the ocean that had never been observed by humans before. By putting enough tags in fish, Lutcavage and Block, like Anderson and hundreds of other academic and fishing-industry taggers, were fleshing out a growing understanding of tuna movements and the ocean temperatures they chased.

In the spring of 1997, Block and her crew took to choppy seas aboard a 10-meter Duffy downeast boat named *Bullfrog*. That day off Hatteras, she had invited Carl Safina along. As part of his role with the Audubon Society, Safina had been following Block's research and was eager for the chance to help her team catch a giant bluefin off North Carolina in the name of science. The week before, she had put some of the very first pop-up satellite tags, those developed by Howey, into a handful of fish. "To the amazement of no one who knows Barbara Block," Safina later wrote, "they were working well." Out on the fishing grounds, Block's team put out the fishing line as the boats surrounding them dropped whole silverfish into the water to draw the giants. Once a tuna hooked on, Safina, who had taken the rod, reeled the hard-fighting fish toward the rear of the boat, where a blue vinyl-covered pad awaited. In their buttercup-yellow waterproof bib pants and boots, Block's team pulled the silver-faced tuna onboard as a big wave helped slosh it onto the deck.

Crouched over the fish, Block and fish biologist Heidi Dewar scrambled to record its length with a measuring tape—key to extrapolating its weight and age. A skinny white hose pumped water into the fish's mouth to keep it alive while its eyes were covered with a dark cloth to keep it calm. Dewar punctured a small hole in the fish's belly, and with a smooth push, Block inserted an archival tag into its abdomen. The most expensive hardware now inside the fish, Dewar took out a surgical stitching kit. With quick jabs and pulls of the surgical thread, she helped sew the five-centimeter slit closed. Taking a wide stride over the huge fish, Dewar punched a wire transmitter and a conventional spaghetti tag into the fish's back, then took a giant step back. Block's team turned the fish to face the water through the boat's deck-level door and, with a mighty tug on the blue pad, projected the fish headfirst back into the ocean. There was a pause before the crew burst into cheering and applause, Block smiling widely beside her crew. All told, the fish had been out of the water for only two minutes.

Howey, whose experience up to that point had primarily been in the field of migratory birds, was surprised when some of Block's tagged bluefin shot across the Atlantic toward the Mediterranean. "Amazingly, [the tags] all worked, but [the bluefin] all swam east and they crossed the designated line of where they're not supposed to go," said Howey. "They all popped up on the other side of the line." That, he said, sparked renewed furor over the management of Atlantic fish stocks.

A few years after the publication of his first book, Safina later accompanied Block's team on another bluefin tagging trip captured on a promotional B-roll filmed for Stanford University. Again their group left from Hatteras, the huge surf splashing onto the boat's front windows as it dropped off the top of a wave, landing onto its valley with a teeth-jarring smack. The first fish was landed smoothly, but on their second tuna they tangled lines with the *Gloria Lee*, out of Boston, and had to maneuver close to disentangle them and ensure their second, even bigger fish wasn't lost. As they tagged that bluefin, the boat's captain struggled to keep the pitching boat steady as Block yelled encouragement to her team. "We need another hand right there," she hollered. "Damn sea, man," one of the crewmembers later shouted. To land a third fish, Safina took to the fighting chair. "Crank, crank, crank," the onlookers chanted as the environmentalist frantically worked the reel. Wearing his trademark ball cap and a green raincoat, one of his shoes patched with duct tape, Safina drew the huge fish toward the boat, where it was eventually hauled aboard. "Come on, godammit," someone hollered as the group yanked the fish out of the water, and Block and her team resumed their tagging choreography. In the cabin, between tagging sessions, she and her team prepared additional tags. "With this technology, we will be able to learn, for the first time, a lot about what these animals really do: how they use their habitat, how they behave—how they *function*," wrote Safina in *Song for the Blue Ocean*. "We will be able to understand them more on their own terms, as though we could pull back the sea's covers and just *watch*."

As Block and Lutcavage continued tagging alongside their funded university programs and interest in bluefin science bloomed, demand for Al Anderson's time and expertise rose in parallel with the rates he charged for a daily charter. As the international fight over bluefin unspooled around him throughout the 1990s, Al had uncharacteristically kept to his own backyard, focusing instead on what he could accomplish locally, evangelizing for bluefin catch-and-release and promoting his chartered tagging trips. He broke his first personal record for tagging bluefin in 1988, when he tagged and released more than 100 flapping juveniles in a single day.

In 1990, less than a year after his brain surgery, Al published two books on bluefin tuna: The first, *The Atlantic Bluefin Tuna: Yesterday, Today, and Tomorrow*, blends oral history interviews with his own memories of the New England bluefin fishery, as well as tidbits and wisdom and teachable moments he had picked up along the way. The second, *To Catch a Tuna*, reads more as a how-to guide for aspiring tuna fishermen. Its cover features a photo of Al wearing orange waders and a blue baseball cap and holding a bluefin nearly as long as he is by the bottom of its throat. A long drip of blood runs down the fish's side, and Al's white teeth are bared in a mix of excitement and effort below his thick mustache and dark aviator sunglasses. "This book, with its theme of how to catch 'em, doesn't mean abandonment of the conservation ethic," Al wrote in that book's second-last chapter as he pivoted to promoting tagging programs. "Just because you have a fish on the line doesn't mean you have to kill it."

After Al stopped drinking, his tagging records and trophies increased dramatically to match his dedication to spreading the gospel of tagging. For 14 years between the late 1980s and 1990s, Al won the Cooperative Game Fish Tagging Program's statuette for most bluefin tagged every year, a number that eventually ran into the thousands. At the same time, he was also setting a growing number of records for tagging thousands of striped bass in the nearby Thames River over the winter season when other types of fishing slowed.

In 1996, Al's close friend Bill Krueger, whom Al had befriended at the University of Rhode Island in the 1960s, asked him to tow the retired academic's old Boston whaler to the dump. The vessel had been sitting on a flat-wheeled trailer in Krueger's driveway for years. "The original wood in the boat, if you could call it that, was badly delaminated," wrote Al, and the old Mercury engine's bolts had rusted in place and had to be hacksawed off. Still, he offered to buy the intact hull, spent around $8,500 refurbishing it, and came away from the experience a happy man with a virtually new boat. The next year, with the smaller boat increasing his catches, he set himself a personal goal to tag and release 10,000 stripers over 10 years—by 2006, 9 years in, he had beat and exceeded that goal, having tagged 15,379 fish, each with its own slip of paper that he dutifully mailed off to the American Littoral Society.

The dreams Al and Daryl harbored for their business had paid off: his bluefin classes and lectures were crammed; fishermen across New England were buying his books and reading his magazine articles; and even though his fishing trips were some of the most expensive out of Snug Harbor, it wasn't unusual for his charter schedule to be packed for the entire bluefin season. Tagging tuna agreed with him, and the trophies that lined his basement haven were a reminder of the value of that work. "Today, more than ever before, researchers need as much information as possible on tuna stocks in the western North Atlantic," he wrote in the penultimate chapter of his first book. "This tagging information can be used to assess populations dynamics, a valuable tool as it becomes more important to manage all tuna species for the future of the fish."

CHAPTER NINE

AMELIA

~~~~~~

Al and Amelia, 2000s

When we try to pick out anything by itself, we find it
hitched to everything else in the Universe.
—JOHN MUIR, *MY FIRST SUMMER IN BOSTON*

The morning of September 27, 2004, Amelia may have been
drawn by the glint of butterfish on the drop as Al's diamond jig drifted down toward her from the *Prowler.* The
lure's movement mimicked a wounded fish, so perhaps that was
what caught her eye. It's also possible that she swam toward it,
lunging at yet another piece of food, another tiny fish to sate her
enormous appetite. Some days she ate as much as 5 percent of her
body weight.

In any case, she struck. On the boat, where Jason Williams and
his brother and father waited and watched, the tip of a rod jerked
violently. A strike from a small fish can feel like an insistent tug,

like a child pulling on a parent's sleeve. A strike from a bluefin, even a small one, is a thunderclap. This fight mirrored all the others that day, but was unique in the way that all fish landings are different. She ran; Williams's father braced himself. She dodged; he took in line. She tired; he pulled her to the surface, her silver skin and bronzed belly glinting in the light. Even when her capture was inevitable, she thrashed her tail at up to 30 times per second, a motion so fast it blurred.

Then the fish-tagging ritual: the fish was netted, unhooked, laid on the scale, and then measured and weighed. That day, Al weighed the fish that would come to be called Amelia at about 10 pounds, which is about the weight of the average house cat. She was the length of about two bowling pins put end to end. Her gills gaped, open and closed, bright-red flesh scraping the air like teeth, searching for oxygen and not finding purchase. Death was near, but Al plunged her overboard with a splash, face-forward. She flicked her tail, a single pound of muscle and sinew powering many times its weight through the open ocean. She dove, stunned but still alive, free again.

Throughout the day, Al festooned his boat's angled outrigger cables with red triangular tag flags, each bearing a blocky white "T" for "tag." Unlike the stiff, lifeless fish brought onto the dock, those fabric scraps boasted of fish caught and released for science, and as Al saw it, his superior moral calling. After landing fish after fish, the fishermen ran out of bait five minutes before noon. By the end of the day, they had run out of tagging flags too. Heading back to port, the boat's red pennants all streamed in the breeze.

Driving the *Prowler* past Point Judith's rocky break, Al smoothly backed into his prized berth. That day, the group had caught, tagged, and released 51 juvenile bluefin tuna, with Jason Williams landing six on his fly rod. They kept one tuna for the group to eat. It had been one of Al's best tuna tagging days ever. That season, Al tagged 1,065 juvenile bluefin tuna—bringing the total number of juvenile bluefin he had tagged by that point to 4,552— more than any single person before or after. But that day, all Al knew was that somewhere deep offshore swam 51 small tuna, indistinguishable from one another except for the yellow scrap of plastic trailing the currents left by their growing bodies.

For three years, Amelia swam alongside her schooling mates, feeding and migrating, leaping and diving. She repeated patterns etched across the ocean by her ancestors. Some days, I imagine, perhaps she sped past Al's glinting hook as the *Prowler*'s shadow darkened the sky above. He had already caught her once; she wouldn't make that mistake again.

In September 2007, Molly Lutcavage headed out onto the waters off Cape Cod with a small team of grad students in a chartered commercial boat to catch and tag bluefin tuna for their own academic research. Once they'd hooked a fish, the team drew each bluefin alongside and into their boat, inserting into the fish's thick, rubbery skin a dart and its tracking tag, attached by a custom tether made by Lutcavage. Inside the device rested a small electronic transponder programmed to communicate with a satellite. For the next year it collected data and eventually sent that data into space, where it provided Lutcavage and her team with a map of where the fish traveled, ate, and possibly even spawned. Studying fish that are dead is one thing; studying fish as they are alive and still swimming is entirely another.

When they pulled in the fish she eventually dubbed Amelia,

Lutcavage was startled to find a weathered spaghetti tag already embedded in her back, near her dorsal fin. For the second time in her life, the female fish had been roughly hauled from the water into the sharp, suffocating air. Now between four and five years old and around four feet long, her oval size slimmed out, she weighed nearly 90 pounds. Working quickly, Lutcavage and her team scribbled down the number on the fish's previous tag before removing it and anchoring a customized PSAT tag of her own to a dart in the fish's back, then tipped the fish, again face-forward, back into the salty sea.

That fall Amelia spent the season foraging for tasty sand lance that darted along Cape Cod's wide continental shelf. Under the water, she fed on fatty bluefish and butterfish and was surveilled by snaggletoothed wolffish as she rocketed past their shadowy lairs. She cruised past the skeletal remains of barnacle-encrusted shipwrecks like the *Heroic*, a 30-meter wooden former US Navy minesweeper that had been repurposed as a fishing vessel and sank in 1969, 24 kilometers off Gloucester, on its way to the Georges Bank fishing grounds. Even at her small size, she could outrun any shark that threatened her school, using a few quick twitches of her muscular tail. When she rested, she had to move at least as far as her own length every second to keep enough water passing across her gills so she could continue to breathe. By mid-November, as the ocean waters cooled, Amelia migrated southward toward the New York Bight, toward the beaches near where Carl Safina had grown up. Along the way, she passed breaching humpbacks and flocks of gulls so thick they seemed to fill the sky.

Returning to her office, Lutcavage reported the fish's recapture, along with the tiny string of numbers on the spaghetti tag, to the National Marine Fisheries Service tagging program. A few weeks later, Al received a slim letter in the mail informing him the fish—recorded as being landed by Richard Williams and tagged by Al—had been re-caught. The letter was exciting but not unusual for Al. By 2007, he had already recaptured dozens of tagged fish, mostly striped bass but a handful of bluefin, fishing in the waters off Block Island. Most of them were his own tags, but sometimes he'd catch a glimpse of another fisherman, another part of the ocean, through the story of a tag that ended up in his mailbox. Any fish that managed to stay alive for decades and grow huge or cross the Atlantic Ocean, that excited him the most, and helped him feel that all the thousands of hours he spent tagging and encouraging others to do the same were paying off. Al also loved the fact that serious scientists like Lutcavage and Block were building on his work.

Between 2005 and 2009, Lutcavage and her graduate students tagged 190 Atlantic bluefin from charter and commercial vessels between Virginia and Massachusetts, 58 with external satellite tags and 132 with implanted tags. For years, the team tracked those fish along the Atlantic coast from the northern Bahamas all the way north into Canadian waters, gathering huge amounts of unique data Lutcavage hoped would help cast light on bluefin behavior. All those discoveries, many hoped, were leading the international community to a new status quo, one in which the market demand for bluefin tuna would finally balance with the long-term survival of the species. But that wasn't to be the case.

In 2008, ICCAT's scientific committee recommended a total allowable catch of 10,000 tonnes of bluefin tuna. Instead, the organization's member delegations chose to set its quota at triple that—nearly 30,000 tonnes, one of the highest quotas ever issued. When added up with the annual estimated illegal catch, the actual global haul was ballparked at about double that, around 60,000 tonnes. That was, as some fisheries scientists calculated it, about one-third of all bluefin tuna left in the entire Atlantic. Environmentalists were furious. "We can't say with certainty that over-exploitation is the smoking gun in the bluefin tuna's disappearance—but clearly there's been a murder," fisheries researcher Brian MacKenzie of the Technical University of Denmark told a journalist in 2007 after he released, alongside his late Dalhousie University colleague Ransom Myers, a study of the collapse of North Atlantic bluefin. He had hoped, he said, that the findings would "inspire a more precautionary approach to the management of bluefin tuna in the Atlantic, with more concern about re-establishing and maintaining the historical range of the species."

Within two years, then NOAA administrator Jane Lubchenco wrote a public statement to ICCAT condemning the organization for its lame-duck oversight of the bluefin. Over 40 years, she reiterated, the organization had "overseen a 72 percent decline in the adult population of the eastern Atlantic and Mediterranean stock of bluefin tuna." The same year, amid global boycotts and encouraged by WWF and other environmental nongovernmental organizations (NGOs), Monaco became the first country in the world to remove bluefin tuna from restaurant menus and retailers' shelves. Capitalizing on global momentum, WWF publicly released a new analysis of ICCAT's own statistics. Those numbers, it claimed, showed that yet again bluefin tuna stocks were close to collapse. In 2009, Prince Albert II of Monaco backed another formal proposal to CITES to list Atlantic bluefin tuna on its Appendix I. For months, as the Netherlands, Sweden, and the UK tried to persuade the European Union to support the proposal, environmentalists thrilled at the prospect that trade restrictions might finally pass. But they knew they were in for a fight.

In March 2010, more than 175 CITES signatories met in Doha, Qatar, for the committee's triennial world conference. Japan, "with bags of money and clever diplomacy, went to work," according to *Tuna Wars'* Steven Adolf. "It couldn't quite be labeled bribery, but in exchange for aid projects in the countries concerned, Japan obtained increasing support to vote against the Monaco initiative." In proceedings the day before the vote, the CITES delegate from Libya took the committee floor and, at the top of his lungs, lambasted the "imperial nations" for depriving his country of its fair share of the bluefin quota and encouraged the gathered nations to scuttle the proposal. (Although he conveniently skirted around the fact that his boss, Libyan dictator Muammar Gaddhafi, had been illegally catching and selling bluefin out of the Mediterranean for years. In one WWF report, it was estimated that 3,600 tonnes of bluefin, almost double the country's ICCAT quota, had been caught in the country's waters in 2005.) On March 18, 2010, the attempted ban on the global bluefin tuna trade proposed by Monaco failed by secret ballot, rejected by a vote of 68–20, with 30 abstentions. It was yet another triumph for commercial industry.

IN AL'S WORLD, it had again turned to a season of loss. In May 2008, after decades of struggling with her mental and physical health, Al's mother, Isabelle, passed away. It was around the same time the New England Marine Fisheries Council[1] allowed net-towing draggers to encroach on his beloved Stellwagen Bank off Cape Cod and, in Al's words, "destroy their wonderful fishing." To Al's mind, the two events felt connected, tragic in their finality. When it came time to bury Isabelle, Al and Daryl wanted to respect her wish to be laid to rest alongside Arthur in New Jersey, so a few weeks later, they traveled south to Fords.

Stopping outside his childhood home on Safran Avenue, Al got out of the car while Daryl waited inside. This, she thought, was a private moment. In the driveway, the home's new owner was washing his own boat with a hose as Al stared in his direction. "Whaddya want?" the owner barked, prompting the captain to approach and introduce himself. Could he see inside his old home, perhaps? Once inside, Al looked around in sadness and wonder at all the things that had changed and others, like the old fireplace with the metal "A" on its outside facing, that were the same. When he got to the kitchen, Al broke down, sobbing, crying like a boy in front of a total stranger. All the pain that preceded that home, all the pain that had been contained within it, and all that followed erupted from him in a flood. And then, as quickly as the feelings had hit, they passed. Wiping his face, Al roughly excused himself, walking out of the home he would never see again before he went to the local graveyard to say goodbye to his last close blood relative.

ON THE EVENING of February 11, 2015, Al and Daryl stepped out of the car in front of the International Game Fish Association's

Fishing Hall of Fame and Museum in Fort Lauderdale, Florida, their faces aglow in the setting sun. For so many decades, Al had aspired to fishing greatness, and that greatness, it seemed, was before him: he was about to join the rarefied IGFA group Legendary Captains and Crew. The year before he had caught, tagged, and released his 60,000th gamefish—a six-pound striped bass—and along with four other captains was set to receive the group's most prestigious award. It was a prize named for Tommy Gifford, the early angler and fishing guide who, along with Michael Lerner, had started that very first east coast rush for Nova Scotia bluefin in Wedgeport nearly a century prior. Many fishermen still believe that Gifford, who died from cancer in the 1970s, was one of the greatest saltwater fishermen of all time.

Al and Daryl took seats at their table inside the museum's blue-lit ballroom, below a firmament of silver-blue fish models that hung from the room's high ceiling. After dinner and a quick slide-show, a fisherman and fellow tagger named Bouncer Smith took the stage to sing Al's praises. "Al Anderson and Frank Mather were the pioneers in what we do today to try to keep our fishery alive," he said, and "Frank couldn't talk a paragraph without mentioning Al." In the decades before tagging existed, back when bluefin were plentiful, it was fishermen like Al Anderson and Frank Mather who were already trying to figure out what made the fish tick, he marveled. As a profession, they had all experienced firsthand how easy it could be, he ended his speech, to wipe out the greatest gamefish in the ocean.

Next up onstage was Paul Osimo, a thick-bearded industrial designer and former mate of Al's who had fished beside him for nearly a decade in the mid-aughts. Back when Osimo was still a boy fishing alongside his dad on Nantucket Sound, he recalled, he'd wander the lure-packed aisles of their local tackle shop. Here, he often gravitated toward a book with a cover featuring a mustachioed man in dark sunglasses holding, as Osimo recalled, "a very surprised-looking, gape-mouthed school tuna." The book—as its title *To Catch a Tuna* suggested—offered insight on how to catch, at the very least, a single bluefin. For years, that book sat near Osimo's bedside; he read it until it was dog-eared. "I will cherish the time that I had jogging 'round the seemingly endless rear deck of the *Prowler*," he said thickly. "Congratulations, Cap."

"This guy's the king," the event's emcee spoke into the microphone as Al carefully stepped onto the stage. Brightly colored embroidered fish adorned the IGFA patch that Daryl had sewn onto his lapel, a riot of fish species standing out against his dark blazer. In the exact middle of his gray tie nestled a shining silver pin of a bluefin tuna. He stood straight at the dais, his hair cropped short, his once black, then gray mustache now white as Daryl gazed at him with pride. His speech was short. "Releasing a tagged gamefish was, and still is, one of the biggest thrills that I've ever had, simply because it offers the possibility, thanks to tagging, of increasing our scientific knowledge about migratory gamefish behavior," he said in his New Jersey lilt. "During Gifford's era, our fishery's resources were quite abundant, whereas the opposite is true today. But I believe that if Gifford were still alive, he, too, would be practicing conservation for science tagging as well." The audience clapped as he walked off the stage, the pyramidal wooden award clenched in his big, wrinkled hands, his eyes slightly moist at the edges as he took his seat, once again, beside the love of his life. Above their graying heads, a painted model of a bluefin tuna swayed in the currents of applause.

CHAPTER TEN

"THEY WILL COME FROM OUTSIDE"

~~~~~~~~~~~~~~~~~~~~~~~~~~~~~~~~~~~

Spain, 2010s

The tuna arrive in great haste, drawn together like a phalanx

of men who march in rank: there are the young, the old, the adults.

And they swim, innumerable, inside the nets.

—OPPIANO OF CILICIA, "ON FISHING"

O n the floor of a bustling Madrid seafood market in the summer of 2018, four huge quarters of giant bluefin rested on a wooden pallet as two police officers hovered above them. One officer was carrying a giant fish-cutting machete while another, in a fuchsia shirt and blue jeans with a gun in a holster on his hip, held a plastic sample container. It was a strange kind of crime scene, but one that unfolded in halls, markets, and transport trucks across Spain and southern Europe on June 26, 2018. Starting that day, Spain's Guardia Civil police force carried out 25 searches and arrested 79 suspects, seizing 80 tonnes of illegal bluefin—part of an illegal market worth about

12.5 million euros annually. (Most of the fish had been caught in Italy and Malta, but a smaller percentage had been caught and hidden in false bottoms of Spanish fishing boats.) Police also found and confiscated 500,000 euros in cash as well as jewelry, watches, and seven luxury vehicles. By the end of the operation, said Madrid-based investigative journalist Marcos García Rey, who covered the story, it was one of the world's largest-ever crackdowns on fishing-related crime.

Guardia Civil officer Escudero (a pseudonym) first started working in his present role at the Nature Protection Service of Spain (SEPRONA) a few months after Operation Tarantelo's so-called action day. His department's investigation had started a year earlier, when the EU Commission first got a tip that frozen tuna intended for canning was being treated with nitrates and carbon monoxide to prevent oxidation and graying of the flesh.[1] The intention was to trick consumers into thinking they were buying fresh tuna—which costs around four times as much. (Preserving fish with carbon monoxide to retain color is legal in the United States but not allowed in the EU.)

The primary concern about illegal bluefin, said Escudero, is the risk it poses to public safety. When stored incorrectly, raw tuna accumulates histamines, naturally occurring chemicals that result when an amino acid called histidine is broken down by bacteria over time at temperatures above zero degrees Celsius. While the human body normally contains naturally occurring histamines that are released in response to allergens, some fish, including tuna, mackerel, and herring, store naturally high amounts in their tissues. If those levels rise even higher and are consumed, the chemicals can cause a severe illness that resembles an intense allergic reaction. It was fear for public safety, paired with the growing sense that bluefin tuna trafficking was reaching egregious levels, that spurred SEPRONA and the International Criminal Police Organization (Interpol) to action, said Escudero.

IN BARBATE, THE annual tuna harvest starts with a sign from the scuba divers, a whistle, and then shouts, as a boatload of men, their forearms uniformly tanned and densely muscled, strain to haul up a giant black net. They stand aboard a large wooden boat painted in thick horizontal stripes of oxblood, gray, and bright blue. Some are young, some old, but all work with a singular focus: they are catching big tuna. On the morning of April 4, 2022, Rafael Márquez Guzmán pulled heavy sections of net alongside the men he managed. Clad in a tight white T-shirt over blue pants, the handsome black-haired man in his 50s yanked the net up with vigor, lifting it with his huge biceps, hooking a section of net onto a winch before grabbing another. As second captain, Guzmán didn't blow the whistle—that was the responsibility of the captain, who used blasts on it to indicate each phase of the coordinated netting and tuna killing operation—but he had worked every position on the almadraba since the early 1990s, back when he was in his early 20s. Now his own 24-year-old son was working on this crew, the fifth generation of Guzmáns to confront and kill giant bluefin tuna on the open ocean off Spain's southern coast: tuna like Amelia, who likely crossed the Atlantic Ocean soon after Lutcavage tagged her, and who lived and spawned in these waters for up to a decade.

In Barbate, the job of working for the local tuna trap historically cascades down the generations. Guzmán's first almadraba took place the year after his father died. When he was growing up, Guzmán's paternal grandfather spun dramatic stories of the perilous, wet work fighting to kill the mighty giant tuna that fueled the economy of this hardscrabble southern Spanish town. Rafa didn't have to become an *almadrabero* like his father; he had already trained as a mechanic. But as soon as he tried it, he knew he was good at it and felt compelled to continue his family's legacy in the fishery. Today he's a local celebrity in his town of about 22,000 residents. Some people gossip that it was a vicious blow from a bluefin's tail that killed his father, a dramatic death to explain the son's heroic and meteoric rise in the fishery. The truth is far more banal: yes, his father had been hit by the tuna a few months before he passed, but it had only cracked a few ribs. But it's a sign of the grandiose reputation that follows these men, these bullfighters of the ocean, along the shore.

Meeting with me in an office at the rear of his automotive garage after a morning spent on the water, Guzmán wore a dark sweater and a tangle of gold and black chains around his neck. When he worked his first almadraba, he felt the weight of his father's respected legacy and that of his grandfather—Rafael—a renowned man in fishing circles whose first name Guzmán inherited. It was once he started scuba diving with the tuna that he really fell in love with the job: the danger, the drama, the teamwork, the feeling of facing a giant fish with only a well-honed blade in hand. On the wall beside his desk, a photo of himself killing a tuna at the almadraba was printed in nearly life-sized scale. In it his muscles, round and bulging, ripple with sweat, and his brow furrows as tiny drops of bluefin blood splatter across the scene. The tuna's head and tail are cropped by the photo's framing, but its flanks, gray and gleaming, suggest the fish had been a worthy match for the Guzmáns' prowess.

Theirs was a primal dance, one dictated for centuries by the structure and timing of the tuna nets. Every summer, Barbate fishermen celebrate their patron saint, Carmen, during an annual festival. They parade with her portrait or a statue, carrying her from the church onto a boat. Thus afloat, she is escorted through the port as devotees line the harbor, their hands clasped in prayer.

The almadraba is a predictable fishery that remains rife with danger for all parties, though it always happens at the inflection point between high and low tide when the current is the calmest. Before tuna enter the almadraba's central killing chamber, or *copo*, they first encounter a net called the *rabera de tierra*, a long expanse of surface-to-seafloor net that runs from the offshore net system until it hits the beach. Tuna swim into this net at a roughly perpendicular angle, then follow its complex interconnected and contiguous series of netted chambers—*palmatorre* to the left or *legitima* and *cuadrillo* to the right—that channel the fish inward toward a one-way entrance called the *boca*, or mouth. Structurally, a bluefin has a large tail and small pectoral fins, so is more comfortable making large looping turns instead of quick turnabouts; it is this tendency that the almadraba exploits. After entering the internal *cámara* chamber, tuna are eventually forced by scuba divers making percussive squeaks from their pressurized oxygen tanks through the *buche* (maw) to enter the copo.

When he started, Rafa worked as a scuba diver on the trap and was responsible for going down into the water with the tuna; it was here he developed a profound admiration for the fish. No matter how fast or chaotically they swam, it was eerie how the fish seemed to see him, how quickly they moved and how they looked him in the eye as they darted past. This would have been Amelia's experience under the water, and I can't help trying to imagine what she was feeling at that moment: perhaps a sense of containment, of panic, of searching for an exit where none existed or ever would again. She would have heard their whistles and shouts, the roars of the motors rattling under the water, the loud pops of the Italian-designed guns called *luparas* that killed the other giant bluefin beside her before she died. She would have smelled and tasted their blood in the water.

When we met in his office in the spring of 2022, Guzmán languidly smoked from a yellow pack of Camels as he recalled how he and the other almadraberos greeted the 2003 quota restrictions dictated by ICCAT and the Spanish government. For decades they had watched numbers of bluefin speeding past their nets dwindle as lawmakers appeared to twiddle their thumbs, and ultimately the announcement was greeted with relief. "The fishery was in danger and we knew it," he told me through a translator, carefully blowing gray smoke away from me in a polite gust. "We were absolutely sure that our jobs would disappear along with the bluefin tuna." He and his fellow fishermen want another 3,000 years of the bluefin fishery in Barbate, he said, tapping his chest with his fingertips, and limiting catches paired with increased demand meant that each fish they landed could be sold for more money—exactly what Al and fishermen like Charlie Donilon had benefited from in the United States. Although, Guzmán added, that wasn't to say that the fishery didn't take a hit. The season before the quotas came in, for instance, they caught 3,500 individual fish. Soon after quotas were imposed, each net could catch only about 600. (That, compared with tens of thousands of bluefin once caught in a single net operated by the 1st Duke of Medina Sidonia hundreds of years earlier, seemed like a drop in a bucket.)

But it was also around the time when quotas were imposed that the fishery seemingly started to get more respect worldwide. Brokers from around the world were coming to Barbate, and even those Spanish people who hadn't known about the almadraba seemed to be showing up for a peek. It was nice, Guzmán said, to feel as if the hard work of centuries, of generations, was finally being recognized and valued.

In 2007, ICCAT implemented its first recovery plan and two years later slashed the total allowable catch to an amount suggested by the scientific advice: from 32,000 tonnes to 13,500. And it was around 10 years later, said Guzmán, that the fishery changed again. Their Japanese buyers and partners concluded that the stress of being dragged from the water by the tail and killed aboard the boats was hurting the quality of the bluefin's meat. Instead, they wanted the fishermen to kill each tuna using a lupara with a single shot to the head. Essentially an exploding cartridge attached to the end of a long pole, the lupara made it easier for the trap's scuba divers to catch and kill each fish un-aware. It was important to kill the fish quickly and humanely, they said, and especially in a way that would get them a better value for each fish. Yet Guzmán immediately felt the loss of that adrenaline-charged moment when he'd face each fish he killed. "In our hearts we feel that it's not the same," he said. Still, the new method was also safer. In previous years it wasn't unusual for al-madraberos to be injured by a hard-flapping tuna; even now, div-ers often leave the net enclosure covered in bruises. Sometimes a tuna's sharp, bright-yellow finlets, which run along its back, rid-dle fishermen's arms with slashes that require stiches.

In the "new" way, once each tuna is landed in the almadraba boat, it must be killed immediately, its spinal column exposed so the fishermen can conduct a procedure developed in Japan called *ikejime*. In the practice, a long, thin rod is inserted down the fish's spine, which essentially stops all the nervous system shocks, lactic acid, and adrenaline coursing through the fish's body. When the fish is too hot or stressed, the resulting heat can damage and even melt the high-value fat within the fish. Once dead, the fish are dropped into a boat hold full of water and ice and transported within the hour to a nearby oceanfront processing plant, operated by the Crespo-family-owned tuna processor and distributor Gadira, where the fish are taken apart and either deep-frozen or stored for refrigerated transport. Since the 1970s and the birth of "flying fish," technology to freeze, transport, and gauge the fish's quality has improved by bounds, including "superfreezers" able to chill the fish at minus 60 degrees Celsius within hours of harvest and expensive three-dimensional scanners to evaluate the quality of frozen tuna flesh.

~~~

The Zahara de los Atunes almadraba, which is easily reached from Barbate's bigger docks slightly to the west, is one of four nearly identical operations that run in Spain's Cádiz region every spring. It is a time of relative moisture, when the baking summer heat has yet to arrive and the hillsides are a psychedelic sprawl of white, purple, yellow, fuchsia, green, and orange. The organization that speaks publicly for the Zahara de los Atunes almadraba, OPP51, also represents the traps in nearby Conil de la Frontera and Tarifa, while another company controls a trap a short ride from Barbate's harbor. The current quota, in 2022, for the three traps is 1,400 tonnes, or about 5,600 individual fish, according to Guzmán. The season's first almadraba takes place after the first full moon in May and always makes the local and national news.

When springtime arrives and bluefin return to their Mediterranean birthplace, thousands of tourists arrive with them, hungry to consume fresh bluefin and the culture that originated these ancient traditions. A short drive away from the town's pale sand beach and network of docks and loading facilities, the front tower of one factory advertises bluefin tours, offering visitors a chance to learn about the fishery's history and witness the butchering of a tuna.

Entering this roadside attraction, which cost me five euros, I walked under bluefin models and nets hanging from the ceiling in moody ocean lighting. Our orderly tour followed a Spanish-speaking guide. She walked alongside a reproduction of a tuna hanging from the ceiling, before suddenly opening a door in its body. She pulled out a huge roe sack longer than a football and hoisted it into the air like a foam finger. Next, she showed the group a cone shape called the *morillo*, a piece of flesh cut from the top of the fish's disembodied head.

Behind wide glass windows, a roughed-up tuna sat atop a slab, its top skin shredded and torn, its fins brown and broken, nothing like any of the prime $1,000 specimens I saw on my journeys from Tokyo to Madrid. This was a tourist fish: a fish with purple-black meat that means it is virtually worthless, there only for the spectacle of its dismemberment. To open the demonstration, a blue-shirted man wearing a headset introduced the fish butchering process, known as the *ronqueo*, or "the snoring of the tuna." Pronounced with a heavily rolled first "r"—*rrrrrrrrronqueo*—the word draws from the zipper-clicking, snoring-like noise the tip of a knife makes as it is pulled across a tuna's top ribs. Roll your tongue, pause, then roll it again, pause, and again: this is the sound a knife makes when scored cleanly against tuna bone.

Before he started, the butcher first gave a shout-out to all the different products that a tuna's body can produce. "Muy bien?" he asked. "Sí!" yelled the crowd. "Perfectamente," murmured the white-haired woman sitting beside me on the stainless-steel bench as she settled in for the show. The room's glass partition muffled the machete's first chop, but the dull, sharp clunks of the butcher's knife still reverberated off the walls. Blood gushed from the fish's body as the worker got to the business of systematically taking the tuna apart, hacking along its pectoral fin, then running the knife along its belly. He moved on to a slightly smaller knife, drawing its blade along each quadrant of the tuna's loins. Within its mottled skin, the tuna's flesh was dark purple where it had been "burned" with lactic acid and heat produced from a stressful landing.

The worker ran his knife along the fish's back, its tip conjuring the strange sound of tearing cloth or a slowly pulled zipper, his breathing controlled but still loud through his amplified headset. The crowd watched, brows clenched, lips pursed, children straight-faced and silent in fascination tinged with horror. (One high-end Barcelona restaurant that offers the ronqueo as a culinary experience was so aware of the shock of the blood from a single tuna that it hired an interior designer to craft a table that tilted away from onlooking diners and tourists—so they could see the blood but were protected from the bulk of its 12 to 15 liters. As their butcher's knife flashed, dark blood would gush, flow, and then trickle toward discreet grates in the floor. On the room's walls, each blue tile was intended to simulate the tuna's scales, its windows framed in red to suggest the fish's most tender inner cavities.)

To conclude, the butcher methodically laid each piece of the fish out on stainless steel shelves for the gathered crowd to ogle, the fish's blood splattering the windows. Bits of flesh dropped off the carcass as the man picked the fish up by its eye socket, dragging it around on the floor to pivot it on the wet ground.

Bluefin is a multimillion-dollar industry in Spain, and Barbate's community has a good sense of what they're selling and to whom they're selling it. In the crowd sat a British couple who live in Portugal's nearby Algarve region. They came to Spain for a few days and were having a tuna feast that week, eating raw and cooked variations of the fish until they weren't sure they wanted to eat any more. The night before, they had dined at El Campero, a white-tablecloth Michelin-starred restaurant with a 95-euro prix-fixe tuna tasting menu. The restaurant mostly serves tourists and is staffed by waiters who wear white shirts emblazoned with a tiny teal tuna over their hearts.

TO UNDERSTAND HOW our era's knowledge of bluefin tracking and physiology fit into the global picture of bluefin, I found myself drawn, time and again, to understanding how one of the world's most ancient bluefin fisheries has grappled with modernity. That search brought me to the doorstep of Ignacio Soto Medina, a man passionate about bluefin, whom I met under a stormy sky in the port of Barbate. He came out of the gate strong and suspicious, asking me what my agenda was, obviously wondering if I was a radical environmentalist. I told him about Al Anderson, about my dad, about Molly Lutcavage and Amelia and my years of trying to understand bluefin tuna and the extent of human power on the future of our planet. He must have been convinced, because he allowed me, along with five Spanish tourists, to board his white, center-console vessel, *Digger*, which carried us to the sea.

Born in the regional hub of Seville, the silver-haired Soto first started spearfishing and scuba diving when he was 14 years old, drawn to the open ocean. He fell in love with open-water ocean swimming, a hobby he still pursues with friends, and eventually bought a second home in Barbate. He has been cannily developing his plan to run high-end tuna tours for more than six years. He'd long aspired to run a business that would share the story of tuna in southern Spain, but it was only in 2022, as the worst of the global COVID-19 pandemic began to ebb, that his plans coalesced. Three of the four almadraba traps on the Cádiz coast have been owned and stewarded by the Crespo family since the early 1970s, and Soto knew the last thing they wanted was a gaggle of tourist boats getting in the way. Securing the Crespo family's permission to bring his boat alongside the high-stakes operation was an early priority and would prove a delicate dance.

First, Soto had to get a prominent member of the family that owned the Zahara de los Atunes almadraba, Diego Crespo, to agree to the partnership. (The Crespos are also part owners of the traps in nearby Conil de la Frontera and Tarifa.) Crespo agreed that Soto could bring tourists out to witness the lifting of the net and the slaughter, but with the caveat that the fleet's captain had the final say on where Soto's boat could park. If he was getting too close, if the weather was too tumultuous to bring the nets together, the captain reserved the right to turn Soto's boat away, even if Soto's tourists were disappointed.

Soto's red-shirted pilot, Luky (real name: Jose Luis), held the license to captain the *Digger*, but it was obvious that the handsome, hazel-eyed Soto was the energetic, charismatic leader of our outing. He made a striking portrait gazing out across the landscape, giant black cows loitering on the beach behind him, the encroaching hillsides flashing by in a blur of red-streaked rocks and fescue.

Beyond just showing tourists the almadraba, Soto said it's important he walk them through the extensive historical, archeological, and culinary elements of the bluefin story in Cádiz. And so he also renovated a run-down fishing hall and built his own slick tourist attraction from scratch. Located in the historic home of the nationalized almadraba fishery—the National Almadraba Consortium—where the Créspos first established their business in 1972, Soto's "working museum" gradually accumulated piles of discarded and overlooked ephemera many in the community considered junk. He scoured archives for rare inked engravings and secured historic photographs from fishermen across the region to better tell the story of Spain's bluefin. In one illustration of Conil from 1564, six workers chop, carry, and salt huge chunks of bluefin, foregrounding a detailed map of the city. High watchtowers, used to signal tuna-catching fishermen on the region's beaches, dot the landscape, with the town's walls and a river sprawling in the distance. Looming above the workers is a giant pyramid of salt towers—back then, they intentionally piled in that shape, said Soto, so they could use early trigonometry to figure out how much salt they had used and how much more they might need. After workers chopped the fish into pieces and salted it in wooden pits, they then packed the preserved fish into wooden barrels, sickled tuna tails littering the ground around their feet. Even until recently, said Soto, wooden barrels were still used to store preserved tuna flesh for months.

To preserve the bluefin's delicate roes, Cádiz's earliest almadraba workers first carefully placed the egg sacs in a vat of water, then pressed them in batches to dehydrate them. For this, they used a huge, rectangular wooden press. Soto recently scavenged the disassembled pieces of that press, cracked and long unused, and assembled it in the museum beside a 100-year-old skiff that would have otherwise been cut into pieces and recycled. Even today, the wooden boats still need to be painstakingly repainted every year, but the dense, heavy boats are stable out on the open ocean and are what the fishermen are accustomed to using.

During Soto's work in getting the museum set up, the Crespo family eventually agreed to partner on the facility, including paying for some of the renovations. Soto hung a giant white flag across one wall of his museum; tuna fear killer whales, he said, and on the water, almadraberos used that fear to their advantage, dropping white flags resembling the bellies of orcas to scare and steer the tuna. Bluefin sometimes even beached themselves in their frenzy to escape the fabric as it flashed in the water below those old almadraba boats.

Gadira, the fish company owned by the Crespo family, also agreed to sell Soto's chef, Carlos Navarro, small tuna cut into cross sections. They keep them frozen until needed for use in live, year-round culinary demonstrations run out of the fancy modern kitchen Soto designed at the museum's end for explicitly that purpose. That's why he hired Navarro, an award-winning sushi chef who was once a finalist in Tokyo's World Sushi Cup, and tasked him with building a menu melding modern Japanese cuisine with those of ancient cultures that prized the fish. (During his research and development, Navarro even worked out a recipe for his own garum, made in the ancient style.)

In 2019, Soto took Chinese billionaire Jack Ma out on a tour of the almadraba fishery after Soto recognized Ma hanging around the docks and offered to bring the Alibaba Group founder out to see the nets in person. The Chinese businessman peppered Soto with questions throughout the tour, fascinated by the process, as Soto later recounted on Facebook. One local Cádiz news site reported on the visit, and in the comments section some locals were wary. "Another who comes to take what is ours," one wrote. "The saying goes: They will come from outside and they will throw you out of your house."

When I visited Barbate, during one week of the 2022 almadraba, I saw evidence of those coming from outside as the Panamanian-flagged freezer vessel *Ocean Star* arrived from Qingdao, China. The newly built vessel is capable of hauling more than 5,000 tonnes of cargo, and it had arrived in town to buy tuna. About 30 percent of the tuna processed and sold by Gadira goes to Japan, said Marta Crespo, Diego's daughter—far less than the 80 percent of global bluefin that the country consumes every year. She also expressed frustration that ICCAT hadn't increased the region's quota, even though, according to what they were seeing in Spanish waters, there were more tuna in the region than ever in recent memory.

Still, with so much still unknown about bluefin, one recent study on multidecadal changes in Atlantic bluefin populations makes the counterpoint to that argument that without deep science proving otherwise, "local increases in abundance should not be used as a reason to relax quotas for commercial or recreational fisheries." But good luck telling that to fishermen and their eager buyers.

IN LATE APRIL 2022, at the Salón Gourmets conference in downtown Madrid, thousands of hospitality businesses, entrepreneurs, and wholesalers were showcasing their wares and trying to rustle up new opportunities. Walking into the multi-hall complex, I headed to the deluxe, carpet-lined booth operated by Ricardo Fuentes e Hijos, one of Spain's largest and most influential bluefin companies and one of its most significant exporters. I approached Isaac Hermo Cespon, the company's commercial director, as he juggled a shifting crowd of suited business partners, journalists, tastemakers, and the occasional curious onlooker. Beyond him, sushi chefs rolled maki and packed nigiri topped with this season's bluefin tuna, fish harvested from the company's Canary Islands operation. Eagle-eyed, Cespon schmoozed and networked, determining how much sushi, wine, and time each table or guests sitting at the booth's bar would receive in accordance with their rank or status, or the level of wining and dining he deemed necessary. Wine was poured, bottle after bottle, as women and men smiled and laughed, raising glasses to one another's health and patting each other on the shoulders. A television journalist and her cameraman set up shop in the middle of the tightly packed kitchen for an interview as sushi chefs worked feverishly behind her.

Their company is the biggest in the Mediterranean, said Cespon, sourcing and selling more than 18,000 tonnes of bluefin tuna every year. (That's the weight of just under two Eiffel Towers.) Fuentes owns facilities in Italy, Morocco, Tunisia, Malta, Portugal, and Spain, and the company has been owned by the Fuentes family since the 1970s, a time when prices for bluefin tuna surged worldwide. While some of its tuna are caught by almadraba, the rest are hauled in by the company's high-capacity purse seiners. Japanese buyers are, by far, Cespon said, the company's biggest customers. Since the 1970s until now, Japan's obsession with maguro has held relatively steady, and it imports about 75 to 80 percent of the global bluefin supply. Still, Fuentes was simultaneously trying to build direct relationships with buyers in new markets, primarily chefs and middlemen in America working in its fish-hungry port cities of Boston, New York, Los Angeles, and Miami. "Before this sushi explosion, tuna was a cheap product— no one was worried about the tuna," Cespon said, later adding that it was cheap primarily because poor handling led to a low-quality product. "Now you can find more sushi bars in Madrid than pizza."

Spain's three largest bluefin tuna companies—Fuentes, Balfégo, and the smaller Cádiz-based OPP51, part of the group that represents the Zahara de los Atunes trap—have long behaved as the reigning kings of bluefin tuna in the Mediterranean, investigative journalist Marcos García Rey told me in an interview at his kitchen table in Madrid in April 2022. "The Fuentes Group has been the largest producer for two decades and boasts of this achievement." They are also famously secretive about their business. So, standing in the Fuentes booth at the culinary conference, I counted myself lucky to be passed off to a chatty high-level official and former bluefin negotiator as he moved smoothly through the nigiri-eating crowd.

In 2003, the official, who requested he not be named for fear of professional retribution, held a high position within a European country's international fisheries department. He was also part of the management team that implemented the first ICCAT quota restrictions on the western Atlantic bluefin fishery, the ones that Guzmán and his fellow almadraberos welcomed with relief. One of the problems with ICCAT, he said, is that the body's scientific committee—which is supposed to be free from political influence—remains haunted by it. Cespon disagreed, saying he sees it another way: after more than a decade, even huge producers like Fuentes have publicly come to accept the necessity of the 2003 ICCAT-imposed cap on catches. "There's a lot of information everywhere, talking about the tuna: 'Oooh, no more tuna in the seas,'" Cespon said, waggling his hands and adding that it's impossible to change everyone's minds about the future of the species. "The problem is not the tuna; the problem was that at that moment . . . there was no regulation [in Europe.]"

Yet later European regulations on eastern Atlantic bluefin catches only capped one part of the problem and inadvertently exacerbated another: specifically, widespread poaching and trafficking of unaccounted-for bluefin tuna out of the Mediterranean. Some estimates held that the illegal network for bluefin was nearly double the size of the legal market. Analyzing data from 2010, the Pew Environment Group issued a report the following year that found the amount of eastern Atlantic bluefin tuna traded globally was about 32,000 metric tonnes, a staggering 141 percent above the legal limit.

While reporting on bluefin tuna with the International Consortium of Investigative Journalists (ICIJ), García Rey traveled across Spain as well as to Malta, Tunisia, Italy, and Brussels. Throughout his travels he interviewed dozens of people, from the CEOs of major bluefin companies to scuba divers responsible for killing and retrieving tuna from the nets. He often worked 14-hour days over the nine months it took to report the story, which the ICIJ titled "Looting the Seas."

During his team's investigation, said García Rey, they found evidence that a small handful of harvesters possibly linked to Fuentes would catch and kill more tuna than legally allowed, using divers to tether the dead tuna to the bottom of the nets in order to smuggle the surplus fish. "It is the most controlled fishery in the world," he said, with a shipboard observer present at every major cull, carefully recording numbers of tuna killed, retrieved, and delivered to the docks that same day. "[So] everything happens under the water." The bodies were then allegedly retrieved later and transported by boat onto shore or aboard other larger ships, a practice referred to in fisheries as "transshipment."

Whenever he goes out for sushi now, García Rey always makes it a point to ask the restaurant to see the tag of the bluefin tuna that he's eating. It embarrasses his wife and kids, but he can't help it. It's the best way, after all, to try to find out if the tuna he's eating—a tuna that, just five years ago, could have been Amelia— is actually legal. Asking the question, he said, is part of educating his children to be responsible consumers and citizens.

After Operation Tarantelo, which was conducted around eight years after García Rey's story was first published, news reports alleged that Fuentes was one of the primary companies that had been linked to the surplus catching and trafficking crimes. In an interview with the industry website IntraFish, the company's deputy director David Martinez categorically denied the accusations. "We do not have any involvement in the activities that they are attributing to us," he was quoted as saying in February 2019. "It is very obvious that some players are trying to involve us in activities allegedly carried out by other companies which, by contrast, are not named in any media reports." When I interviewed Cespon at the Fuentes booth in 2022, he had brushed off the prospect that anything would happen with the case against the company in the country's National High Court. "It's been four years and nothing has happened," he scoffed. "They don't have a case."

One of the primary purposes of the SEPRONA operation, Escudero noted, was to reassure consumers across Europe and the world that Spanish bluefin was safe, and that people would continue to eat tuna and feel good about it. Working for SEPRONA on behalf of consumers and the environment, he pointed out, is a special calling, and one that requires specifically complex work that spans a huge range of legislative departments and topics. In drug trafficking, for instance, two pounds of cocaine is simply two pounds of cocaine. But in wildlife or environmental crime, each investigation requires its own complex balance of proofs and science, ranging from necropsies performed on tuna by veterinarians, to specialized laboratory tests, to chain-of-custody documents with constant refrigeration and protection of samples along the way.

The day after my meeting at SEPRONA, I scored a Saturday lunchtime seating reservation at Madrid's Kappo, a Japanese-style sushi bar opened by chef Mario Payán in 2017. The day before my meal, Payán had driven to the fish market at Mercamadrid to buy enough fresh fish and shellfish to supply his Michelin-starred restaurant to serve diners over a busy long weekend. Like his staff, Payán wore a stone-gray apron over a stone-gray shirt as bottles of sake and white wine were gracefully poured by smooth-moving serving staff. He worked behind the bar, omakase-style, rolling and packing nigiri sushi with fluid, performative grace. A brown lobster sat in a brown-and-red Japanese bowl, a silver throwing star embedded in the wooden column above it. A bearded man in jeans and a jean shirt sitting beside me ordered an expensive white wine as Payán prepared and passed nigiri, one by one, across the counter, dropping each onto a teal-and-black ceramic block. He patted them together and placed them on the block casually yet intentionally, each flick of the finger practiced, each soy-sauced brush of raw fish mesmerizingly lethargic but also weirdly efficient. Copper-clad wooden pots of lukewarm rice rested beside both Payán and the sous-chef to his right as they gently sawed their razor-sharp knives against each piece of fish to enhance the texture of each species: bonito tuna, trout, yellowtail tuna, grouper, mackerel, parrotfish, swordfish, squid, surf clam.

Every piece of sushi received a pat of wasabi or smear of soy sauce with the brush according to a well-choreographed dance, a shiso leaf pounded between fist and palm before being added to a dish. The red mullet wore a cap of sea urchin, right before Payán served tuna belly atop a scallop slice. Then, in a crescendo, Payán smoothly delivered three sleekly cut pieces of almadraba-harvested tuna—otoro, mid-belly, and loin—that had been caught a week prior in his hometown of Barbate and aged for five days in the fridge using the Japanese-developed method. (The man in the jean shirt ordered caviar on one tuna nigiri in a fit of decadence.) At the end of the service, Payán broke down a whole tranche of tuna, getting cuts ready for the evening service. He held the red-pink loin toward me to show it off, its fat glistening, its weight more expensive, per pound, than nearly any other seafood on the planet.

CHAPTER ELEVEN

BEGINNING'S END

~~~~~~~~~~~~~~~~

Al and Amelia, 2010s

Time is but the stream I go a-fishing in. I drink at it; but while
I drink I see the sandy bottom and detect how shallow it is.
—HENRY DAVID THOREAU, *WALDEN*

In the late spring of 2018, Amelia entered the warming waters of the Mediterranean to spawn. Now weighing about 600 pounds, her belly bulged with a packed load of around 35 million eggs, each about a millimeter wide. Once the water reached a temperature of 20 degrees Celsius, it was finally time. Under a waxing moon at around 2 A.M., Amelia joined a school of mature adults as they began to swim in a huge circle. Suddenly, she lunged toward the surface and corkscrewed her body sideways as she released a spurt of eggs into the path of an oncoming male. The skin of their bellies touched as he responded with a burst of milt, a sperm-laced white cloud that grew and spread underwater with each mating, one that allowed fishermen to track the school from the surface.

Moonlight flashed off her silver belly, and by the end of the spawning week, she had lost between 15 to 20 pounds in egg weight alone. There wasn't much to eat in the waters around her: it was time to head back to the Atlantic Ocean before she starved.

And so, in the summer of 2018, a skinny, weak Amelia traveled westward out of the Mediterranean within a kilometer or two of the shore. As she swam past the shallow, sandy island marshes flanking the Portuguese town of Olhão, she encountered a complicated architecture of deep vertical setnets, owned and operated by Portuguese employees of Tunipex. Those nets were imported from Japan and stationed in a prime tuna migration corridor a short boat ride away from Olhão's port. The next, short phase of her life would be spent in captivity.

For three months she was held within a deep ocean pen and fattened on high-quality mackerel and other small fish caught by Tunipex earlier that season, until two-meter-long Amelia weighed 642 pounds. On the morning of October 25, 2018, just offshore from Olhão's bustling daily seafood market, a Tunipex-paid scuba diver jumped into the setnet's final chamber. He lined up his weapon, watching the dark-silver fish swim her final loop. He held his deadly pole steady, hovering in the water as he waited for his moment. Then a muffled shot thumped out from the end of his lupara. *Pop.*

Amelia died instantly from a direct shot to the head, her blood lazily billowing through the water as her eyes grew flat and empty.

After her death, and the death of dozens of other tuna that morning, came a frenzy of activity. The best value for tuna comes from the freshest fish; with every minute that passes, the fish loses value. Before scrambling back aboard the boats, scuba divers attached ropes to each tuna's tail, and workers in rubber overalls used a hydraulic winch to lift the fish onto their low-hulled skiff.

Soon after Amelia's body was hauled aboard, Alfredo Poço, the trap manager for the Japanese-Portuguese company, noticed a 10-centimeter long piece of plastic embedded in her back. The tube resembled a discarded glow stick, and at first Poço thought it might be a parasite. But within its hollow center nestled the spaghetti-like tag, which, upon closer examination, he saw was embossed with tiny black numbers and a word in English—"REWARD."

Poço quickly called Portuguese government biologist Pedro Lino with news of his strange discovery. Lino, a swarthy, square-jawed tuna expert who works in Olhão at the Portuguese Institute for the Ocean and Atmosphere, asked Poço if the fish's head had already been discarded: if saved, an ear bone within it called the otolith could provide valuable information, including how old the fish was and where it had been born. To his disappointment, it had already been thrown away. Within minutes of Amelia's death, the fish had been gutted, cooled, and shipped to Madrid, where she had likely ended up at a high-end restaurant, with her rotting head in a waste bin outside the market. He emailed a contact at the United States' National Marine Fisheries Service, curious about the twisted, weathered tag. Who had tagged that fish? And, more important, what secrets did its tiny string of numbers contain?

BY THE DAY Molly Lutcavage caught wind of Amelia's recapture and death, her professional career had been gradually losing steam. Years' worth of congressional appropriation funds that had supported her Large Pelagics Research Center, which she had founded at the University of New Hampshire in 2003, had expired. She had long become accustomed to the fickle ebbs and flows of university research funding, but she had never been tenure-track and this newest set of losses felt existential. In 2011 she found a new home for her LPRC lab at the environmental school at the University of Massachusetts Amherst, and eventually at the school's Boston campus in 2015, where students interested in working with tuna could split their time between her Gloucester headquarters and the school's harbor campus. Over the years, Lutcavage downsized her Gloucester lab three times; the most recent office space she had found to rent was in the attic of Maritime Gloucester, a museum and aquarium on the city's harbor.

Despite working with less money and no graduate students, she continued her research and collaborations, heading out onto the ocean with commercial tuna fishermen to tag tuna—including in Wedgeport, Nova Scotia—and taking and analyzing samples of tuna flesh and organs at every opportunity while collaborating with other tuna researchers. In early March 2016, she and a group of collaborators published an exciting new paper in the well-regarded journal *Proceedings of the National Academy of Sciences*. Led by primary author and NOAA scientist David Richardson, the paper heralded the important discovery of a bluefin tuna spawning ground in the Slope Sea, a patch of ocean sandwiched between warm Gulf Stream waters and the sloped drop-off just beyond the US continental shelf. It also published pictures of scrawny bluefin larvae that had been gathered in water samples.

Lutcavage knew Frank Mather always suspected that the old-est, biggest western bluefin spawned first in the Gulf of Mexico, with younger, smaller fish spawning farther north, near Nova Sco-tia, later in the season: a shift that made the species less vulner-able to overfishing. Some old-timers even went so far as to say that Mather's theory was suppressed because it wasn't conve-nient for fisheries managers to incorporate into their models. "My Japanese scientific colleagues would sit next to me at meetings at ICCAT and just roll their eyes," she says of the long-term denial by environmentalists of the possibility of western Atlantic spawn-ing grounds outside the Gulf of Mexico. Those Japanese experts, she says, would show her old maps indicating exactly where they had caught the fish historically that reinforced her theories. "If the Japanese experts on bluefin biology tell me that this is the way it really is, I'm going to listen to them," she says. "They're really good at bluefin science."

In a responding letter published in July 2016, Carl Safina blasted the Richardson paper, pointing out that the authors' data didn't support the paper's findings that the Slope Sea spawning area was a major new discovery, or that western Atlantic bluefin spawn younger than previously known and were therefore less vulner-able to fishing pressure. Furious with Safina, her old rival, Lut-cavage responded with an excoriating public online post on Medium that was quickly shared and quoted on commercial fish-ing blogs and across her social media circles.

In it, Lutcavage slammed Safina as a sexist "environmental bully," detailing their first encounter, which she said was the con-flict they had over the aerial fishing surveys she ran for the New England Aquarium in the early 1990s. "To claim to be an expert where you are not, to mislead the public, to falsely disparage those that don't support your ideology, to repeatedly and falsely allude to a woman scientist being bought by fishermen . . . Is this what conservation leadership has become?" she wrote. "Our scientific research got in the way of the extinction agenda and continues to do so today." In the post, Lutcavage dredged up a screenshot of a six-year-old blog post that once appeared on Safina's blog: "Lutcavage has spent much of her career pie-in-the-skying about unknown holy grail spawning areas," he had written. "The fishing industry, with which she enjoys a cozy relationship, would love it if there were. That's because they think it would help their denial that the bluefin is really as depleted as every other independent academic and government scientist and even the Atlantic tuna commission say it is." This echoed his alleged attack from decades before: that he believed Lutcavage's proximity to commercial fishers meant her science should be considered suspect.

Then Safina went a step further, posting an article entitled "Bluefin Tuna: New Study Doesn't Hold Water" on his own Medium page—a page that, as of August 2022, had nearly 2,000 followers, compared with the 53 on Lutcavage's page. In the 3,000-word essay, Safina broke down the Richardson paper's Slope Sea spawning assertions point by point, interspersing his own photos of bloody, dead tuna with captions including "Killed to make sushi" and "Butchering, part II." In that piece, Lutcavage's name comes up only in the references section, but in a separate post that included a seven-bullet-point list, Safina responded to Lutcavage's allegations, which he called "a lot of empty lashing out." He pointed out that she had referenced "a 25 year old letter she can't find" and went out of his way to maintain that he was "in fact a proud feminist." He also recalled his first encounter with Lutcavage, which he said occurred at a scientific meeting where, "inexplicably, she wouldn't look at me or say hello."

What had ostensibly started as a scientific and moral disagreement thus ballooned into a messy public feud. Notably, though, if you asked either one of them, as I later did, they would both insist that their intention was simply to understand bluefin tuna better. In Safina's case, he also proudly underscored his role in trying to protect the fish; in Lutcavage's case, she insisted that the data clearly showed the fish didn't need protection at the level Safina had long claimed. Lutcavage and Safina also appeared to have more in common than I'm sure either would want to admit: they're both smart, congenial, and deeply admired by their colleagues; both work hard and are leading experts in their fields. It felt strange to admit to myself the possibility that, even after spending years learning about their rationales and their bitter feud, I'd never know who was more right, or whose point of view history or future tagging work would prove more correct. Science exists in suspension with human nature, and I found myself deeply split.

AS PEDRO LINO wondered about the origin of that weathered yellow tuna tag, he eventually hit pay dirt. He had sent a report of the tag's recapture—tag R197521—to staff at America's Southeast Fisheries Science Center in Miami and forwarded all the data he had, including where the trap-fattened fish had been caught

(37.01332° north, 7.71035° west), her sex (female), her capture date (between June 24 and July 31, 2018), and her length from the tip of her nose to the middle of her tail (238 centimeters). They, in turn, tracked down Molly Lutcavage and confirmed that the fish was indeed one with a remarkable four data points recorded over a 14-year period: the first was Al's initial tag; the second, the deployment of Lutcavage's pop-up tag; the third, when that tag released and transmitted its data; the fourth, the fish's final, thrashing moments in the Tunipex setnet. They also had the data from the satellite tag Amelia had carried for a full year in the western Atlantic.

"If I understand correctly this fish was tagged twice in his lifetime?" wrote Lino in an email to Lutcavage and her postdoctoral fellow (now collaborator) Chi Hin Lam. "That is quite unlucky for him :) but interesting for us. It is indeed a shame that we did not get a chance to collect tissue or age structures for this fish." Even without that data, Lutcavage's team was excited by the news. "We are extremely thrilled our tether withstood 11 years of sea abuse and contributed one of the (few dozens) of longest tag recovery records of any pelagic fish," wrote Lam, who mentioned that he and Lutcavage wanted to present the data or write a paper on their recaptured fish but that the "n"—or sample size—was too small for most scientific journals. But they trumpeted the news of the tag's recovery on the group's Facebook page, and eventually built Amelia her own website.

Yet all the good data in the world couldn't save Lutcavage's Gloucester research lab. In 2014, she lost her final tranche of federal funding and was later forced to move all her equipment, materials, and samples into her home garage when the building housing her laboratory was sold. But she refused to stop working on her life's passion. Speaking with *Boston Globe* reporter Billy Baker about a newly reported wave of mysteriously skinny tuna that had arrived off the Massachusetts coast in 2019, Lutcavage gave him a tour of her new operation. "This is now the Large Pelagic Research Center," she announced to him in a voice, he wrote, that sounded like she was "on the fence about laughing or crying." Within the garage, he observed, files, lab equipment, and boxes of microscope slides "crammed in alongside gardening equipment and winter coats and her surfboard." She and Lam had carted all the LPRC's remaining equipment and supplies out of the museum attic only a week before. "Without university support and with her grants all dried up, she's able to pay herself and Lam only a fraction of their salaries, mostly from contract work," Baker wrote. "And because universities expect researchers to bring in grant money, she fears UMass will soon pull their affiliation entirely."

One day, Lutcavage said, a friend of hers commented that it seemed like her job had turned into her hobby. Baker, during his interview, asked Lutcavage if that was true. "I'm committed to the unknown," she replied. "The scientific mysteries that no one has solved will always continue to drive me." And it took nearly another decade until one of her career's most major discoveries proved to be correct.

In 2022, a young marine scientist named Christina Hernández copublished a study of bluefin larvae scooped out of the western Atlantic that corroborated Lutcavage's findings of a potentially major new spawning ground for tuna from eight years prior. Hernández, who is currently working in a postdoctoral position at Cornell University, still remembers the first time she read the controversial 2016 Slope Sea paper that Lutcavage coauthored. It was back when she was in the first year of a joint MIT and Woods Hole PhD program. "I was like, 'Oh my goodness, there are things still to discover about the ocean,'" Hernández told me. "I just went to my supervisor and said, 'This is really cool. I'm excited about this.'" So Hernández embarked on a follow-up project analyzing a new round of tuna larvae collected in 2016 for her doctoral degree.

From 2016 to 2020, Hernández worked on the tiny bluefin larvae, working with dissecting pins to extract microscopic otolith ear bones from their wafer-thin bodies. "They're smaller than a grain of sand, completely invisible to the naked eye," she said. After drawing a circle on a slide with a permanent marker—so she could find it under a microscope—she mounted each otolith in immersion oil. "It was awful. At the beginning I would rip the head apart, and I wouldn't be able to find the otoliths." After looking for the visual pattern of each calcium carbonate structure, she found evidence that reinforced the findings of the Richardson group and, by extension, Lutcavage's work. Hernández's findings—evidence that the Slope Sea is indeed a major spawning ground for bluefin tuna, potentially comparable in numbers with the Gulf—made headlines across North America. It also reinforced Hernández's feeling that Lutcavage had been unfairly treated by the scientific establishment. "It makes me really sad to think about that she's finishing up her career really, really fighting to work," Hernández said. "Barb[ara Block] and Molly are fascinating people, and they're both very strong women, and it's a shame that they couldn't be each other's cheerleaders . . . But the social dynamics, especially for their generation, were really, really hard."

Throughout the aughts, teens, and early 2020s, an increasing wave of science has undermined the validity of the two-stock theory, publicly pulling the rug out from under a once-influential idea that has been quietly disdained by scientists for decades. In 2022, Canadian fisheries biologist Gregory Puncher copublished a study on fin clips and muscle tissue of bluefin tuna caught in western Atlantic waters between 2004 and 2018, including some caught in Wedgeport, Nova Scotia. In it, the authors determined that, of the samples they could definitively identify, more of those bluefin had been born in the Mediterranean than in the western Atlantic, although they couldn't pinpoint the origin of a large percentage, about 25 percent, of fish. (Some researchers have suggested that those fish may actually be bluefin originating from an as-yet-undiscovered new breeding ground or a genetic hybrid of both western and eastern bluefin.)

It's a body of evidence that continues to grow. In the 2019 book *The Future of Bluefin Tunas*, edited by Barbara Block, the French marine ecologist Jean-Marc Fromentin and his coauthors wrote an article about the "important new information" that electronic tagging has revealed. "Alternative hypotheses . . . might require adaptation of the management regime," they wrote, and "failing to do so might cause future overexploitation." The same year, Antonio Di Natale, a former coordinator of ICCAT's bluefin tuna program, published a paper asking the scientific community if it was time to officially retire the two-stock theory. "There is only one Atlantic bluefin tuna stock," he wrote. "This new (and old) scientific approach should be duly presented and explained to the ICCAT Commission, knowing . . . that it would create hard discussions."

Theories of bluefin spawning aren't the only ones that could soon topple. Aided by high-tech computer programming languages, scientists believe they are close to replacing fisheries' long-derided maximum sustainable yield with new management frameworks that will allow fish populations and our societies' food systems more wiggle room. It is wiggle room that I believe we must have in the face of the unprecedented human-caused climatic changes currently occurring on our planet. The jargony term "maximum sustainable yield" may be deceptively benign, but in my opinion, those three words and their acronym, MSY, have done a century of harm. They encapsulate that problematic vision of humanity's right to control the ocean once correctly dubbed "policy disguised as science." It is this same problematic vision that allowed global "merchants of doubt"—another term wielded by Naomi Oreskes— to buy corporations the chance to make money at the cost of our planet's atmosphere and stable climate. How dare we use science as a smokescreen, when instead what we are defending is our own unjustifiable appetites and human hubris. If science is indeed the purest form of knowledge, as my father tried to teach me, then why does it often feel so corrupted?

Yet even in my darkest moments working on this book, the hope and positivity of scientists working in the bluefin space, particularly young scientists who have spent their careers working for environmental NGOs in an effort to help protect bluefin, helped me push back against the cynicism I often found overwhelming. One of the brightest lights, I felt, includes Pew Foundation biologist and conservationist Shana Miller, a former graduate student of Barbara Block's who has earned herself a sterling reputation in both commercial and environmental quarters. (She caught her first tuna, a 154-pounder, in 1998 with some friends off Maryland's coastline; not knowing the fish was overfished, they ate it.)

Overthrowing MSY's slim margin for error is increasingly important, Miller says, particularly as uncertainties of how climate change will affect our oceans and bluefin populations come barreling toward us. When we first talked, she was particularly enthusiastic about one newer approach, called harvest strategies or management procedures, that uses a science-first framework to set a long-term vision for achieving and maintaining the health of fish populations. A management procedure has been in development for Atlantic bluefin at ICCAT since 2014. They are akin, she says, to establishing clear rules before a game is played: rules that must be followed even when unforeseen circumstances arise. The approach has already worked in helping Southern bluefin stocks recover to 20 percent of their estimated pre-fished levels. In the same region, fisheries managers were able to increase quotas by 50 percent as fish populations rebounded. "The managers still have an important role," says Miller, "but the science isn't as open to tampering."

In British Columbia, quantitative marine ecologist and University of British Columbia associate professor Tom Carruthers has become a global leader in a complex computer-fueled mathematical model called management strategy evaluation (MSE). Many, including Miller, have long hoped the model, which supports the development of management procedures, could serve as the scaffolding for a new, more accurate framework for managing bluefin tuna in the Atlantic. "If someone says 'I don't like that model estimate,' the first thing you say is, 'Well, what information data are you privy to that is not already in this model?'" Carruthers once said in an online demonstration of his work, "If you don't have it, then what's your basis? What's your proof?" MSE incorporates myriad data sets including population models, otoliths, oceanographic statistics, and catch landings to more accurately extrapolate how bluefin populations are faring. It's the first time such a wide range of electronic tagging data have been incorporated into a model, which promises to give fisheries managers a clearer, long-term picture of the species. "Barbara's tags, Molly's tags, the NMFS tags, European tags, everybody's tags are in [the MSE model]," says Miller. While conventional spaghetti-tag returns like Al's aren't included, his work set the stage for the electronic tagging work that followed. "It definitely was the foundation," she says.

In November 2022, Miller joined hundreds of delegates from around the world in Vale de Lobo, Portugal, for ICCAT's annual meeting, where a management procedure for Atlantic bluefin was on the agenda. She spent a tense week watching proposals and counterproposals being tabled and debated as some countries' delegates and industry lobbyists tried to water down the plan or scuttle it entirely. It took solidarity among Canada, the US, and North African countries, including Libya and Syria, says Miller, but the negotiations resulted in a three-year management procedure agreement for the species that ended up passing to cheers.

Miller and a few of her fellow delegates teared up as the measure was adopted. "We are returning home with a package of decisions that moves us into a modern fisheries paradigm that takes climate into account for some of the most iconic and sought-after species in the world," declared US commissioner to ICCAT Kelly Kryc in the meeting's wake. Miller echoed her sense of relief and possibility. "Everybody's tired of fighting every year about bluefin," she told me the day after her flight back from the meeting. "This is the way fisheries management should work." With the Atlantic bluefin "safe and secure" for the foreseeable future, Miller, who has worked on bluefin for most of her working life, joked that she needs a new hobby. It also means that if you're at a sushi restaurant and Atlantic bluefin is on the menu, she said, you can order it without hesitation. That, to me, feels profoundly hopeful.

At a governmental level, attitudes on the importance of protecting fish populations from collapse also appear to be changing. "Long story short, it just took time," says Boris Worm, a marine ecologist and coauthor of two influential fisheries conservation papers that, looking back, were seminal for a field that is still rapidly growing and evolving. In 2006, Worm and his longtime friend and collaborator, Ransom Myers, copublished a paper that suggested that nearly 90 percent of global fish stocks, including bluefin, could collapse by 2048 if industrial fishing continued apace. When Worm first ran those numbers, while overseeing a group of students taking an exam at Halifax's Dalhousie University, the hair on the back of his neck prickled. "This cannot be true," he thought before he ran the numbers again. He then double-checked them by hand. An article about the paper's findings ran in the *New York Times* and grabbed headlines—and garnered attacks and accusations of sensationalistic hyperbole—across the world.

The second paper, authored in conjunction with respected University of Washington fisheries expert Ray Hilborn and 19 other researchers, essentially formulated a recovery plan for the two-thirds of global fish species that were, at the time, overfished. The paper took two years of collaboration and countless hours of collecting and collating data from across the globe, but perhaps its most significant accomplishment was in uniting Worm and Hilborn. Three years earlier, Hilborn had publicly spoken out against Worm's 2006 paper, in a move, one writer said, that "was like Superman calling out Batman." But during a call-in show on NPR, the pair found themselves talking about the type of scientific meta-study that might show the way forward and, astonishingly for both of them, found that they had finally agreed on something.

As Worm saw it, that second paper, which was published in the July 31, 2009, issue of *Science*, presented a formula that was practical and possible: put stock assessments in place for commercially caught species, follow advice by experts when reductions or tweaks are needed, and customize approaches to regional communities. Examining efforts to stop overfishing in 10 large marine ecosystems, the paper identified 5 where those efforts were working and extrapolated lessons for the rest of the world. "We envision a seascape where the rebuilding, conservation, and sustainable use of marine resources become unifying themes for science, management, and society," the authors wrote. "The road to recovery is not always simple and not without short-term costs. Yet, it remains our only option for insuring fisheries and marine eco-systems against further depletion and collapse."

In that 2009 paper, Worm, Hilborn, and their fellow coauthors laid out an action plan; they had showed how simple the problem was and how it could be fixed, and now all they needed was con-crete action from government to implement a suite of policies that would, as Worm saw it, be a win-win-win for everyone. The Can-adian government found his work interesting, he says, but he left a meeting with the country's fisheries minister with no outcomes, follow-up meetings, or commitments. Similar US meetings with members of Congress and other politicians led to similar out-comes: neat paper, they told Worm and Hilborn, but we're not sure what you expect us to do about it.

"I'm impatient," Worm said from his home in early 2022. "I thought [something] would happen the next year, it didn't, and I got frustrated." He didn't know what to do next. They had done something that no one thought was possible, and still it seemed to have been worth nothing.

Yet in the years following the paper's findings, he noticed devel-opments that would have been largely unheard-of decades before: new collaborations between respected scientists and environ-mental NGOs became more common, while commercial fishers seemed more publicly open to embrace catch reductions, harvest strategies, and lower quotas. ICCAT itself also seemed to im-prove. "Whereas in former days the Commission developed a reputation for finding ways to work around the scientific advice, it appears to have successfully arrested overfishing," one 2019 paper observed.

At the same time, programs like the Marine Stewardship Council's seafood certification program took off and gained purchase as populations of bluefin started to show hopeful recoveries all around the world. All that work, combining public lobbying by NGOs, increasingly better science and data collection, and concessions from tuna-fishing nations, meant bluefin started popping up in waters where they hadn't schooled in years. At the same ICCAT meeting where delegates adopted new harvest strategies for Atlantic bluefin tuna, they also approved higher catch levels in the eastern Atlantic, which means fishing operations like Guzmán's will be able to catch more fish.

It helps that the mood of the debate around bluefin has gotten "massively better," says tuna researcher Laurenne Schiller, who cites Safina's *Song for the Blue Ocean* as having a formative influence on her own career. She started following news about the fish closely in 2008, the year, she recalls, Greenpeace activists dumped a pile of dead bluefin heads on the doorstep of France's fisheries ministry to coincide with that year's ICCAT meeting in Marrakech. "My impression is that the science is much more listened to than it was in the past because more people are paying attention," she says. "This issue is now at the forefront of people's minds and there is more cohesion between NGOs and industry now too, which makes a huge difference."

There's also additional research suggesting that bluefin's incredible physiology may help protect it—at least for a short window of time—during our collective uncertain future. "Bluefin is one of the rare fishes I have seen so far that seems to be positively affected by climate change and global warming," says Fromentin. In 2003, a massive heat wave swept Europe and warmed the Mediterranean, but instead of hurting bluefin populations, it had the opposite effect. Bluefin larvae need waters of at least 20 to 25 degrees Celsius to survive, and the cohort of young fish from 2003, he says, appear to have doubled that year. "Maybe it's another story in the Gulf of Mexico."

That does indeed appear to be the case. In 2011, fisheries oceanographer Barbara Muhling and her collaborators published a paper with worrisome findings for the future of bluefin spawning in the important western spawning zone. Tagged bluefin generally avoid waters hotter than 30 degrees Celsius, and three decades of plankton surveys by US government scientists have never found bluefin larvae in Gulf of Mexico waters warmer than 29 degrees Celsius. Yet under current climate change scenarios, the already temperate waters of the Gulf are set to warm far past that point. Under current mean global temperature predictions, Muhling concluded, the areas of the Gulf of Mexico where bluefin currently spawn in the late spring would be reduced by about 40 to 60 percent by 2050, and by 93 to 96 percent by 2100. It's possible that spawning could be pushed to other zones, like the Slope Sea, or to other times of year, like the early spring. But the realities of the species mean it's "likely to be vulnerable to climate change." I'm inclined to believe that's likely true for bluefin no matter where in the world they currently are; if a species cannot reproduce, it is doomed to perish.

In the spring of 2022, I was sitting in a giant boardroom in Olhão, my hands shaking as I reached across a table. Pedro Lino, the scientist who had initially been called with the report of the tag embedded in Amelia's side, had just presented me with a plastic sample vial. Within it rested the yellow plastic tag that Lutcavage had told me about years earlier. This was her spaghetti tag, the remnant of the pop-up tag that Lutcavage had meticulously embedded in the heart of the PSAT device that had popped off years earlier. Right after the scrap had been extracted from underneath Amelia's skin, a Tunipex employee had soaked it in bleach to remove the smell. Even after she dropped it off at Lino's office, it stank. Now, four years later, I held it up to my own nose and . . . nothing. Even that fleeting remnant of Amelia had vanished.

Lino likes the "Molly Lutcavage" school of tagging, he said, because her group has long used medical-grade materials, like silicone, in their tags and seems to take great care while implanting each one. Growing up in Portugal, Lino—like Lutcavage and Morales-Muñiz—loved Jacques Cousteau's television programs, which inspired him to study marine biology in school. Unlike the other researchers, though, he doesn't define himself as a man obsessed with bluefin. "Because bluefin tuna is in such a good state, I'm not worried about bluefin tuna at all," he said. In Portugal's Algarve, unlike in Spain's Cádiz region, it's nearly impossible to find local bluefin tuna because essentially all of it is exported now. "All fisheries are about human nature. It's about catching fish," he said. "If you work with fishermen, you know that most of them will catch whatever there is in the water, because they're not sure if tomorrow they will catch these fish. And if they leave it, someone else will catch it."

When I pointed out that approach can be the start of a literal race to the bottom, Lino didn't disagree. "My personal opinion about this is that we need to catch less, eat less, and make the most of what we catch," he said. "You go to a restaurant and you eat an obnoxious amount of food that you don't need. Then you feel too full and then you will have digestive drinks so you feel better. And then the next day you will go to the gym because of all that food that you ate." Blaming fishermen is easy, he said; solving the problem of how humanity catches and consumes our planet's resources is a far more urgent question.

Science doesn't hold abstract authority. It is a practice and a profession incrementally built by trial and error over generations, and I am always wary when any one person or institution claims power or authority over its gradual, deeply human process. When ICCAT drew that 45° line down the Atlantic, its member countries acted out of political expedience, not out of scientific certainty. But over resulting decades, when challenged, they defended that line and ICCAT's policies as a scientific truth. These are the shifting baselines of history, the tempting supremacy of groupthink currently undermining today's ongoing scientific progress. Perhaps worse than junk science, I feel, is science divorced from context, and particularly science weaponized in service of the mindless capitalist push toward progress at any environmental or human cost.

After spending years of my own life chasing bluefin around the world, I've ultimately come to believe that as a global community, we are collectively only ever a few terrible choices away from wiping out any ocean species. That's especially the case if we're not paying close enough attention to the fish we eat, where we're buying those fish from, and how those fish are managed, caught, distributed, and sold to us. The bluefin tuna's astronomical value also makes fishermen especially vulnerable to greed, that central human vice. In the absence of meaningful public policies limiting catches, fishermen will simply catch more fish than our oceans can support, and species will collapse in an unpredictable trail of dominos. And that's *even though* most savvy fishermen know that catching too many fish can tank market prices, and that the long-term consequences of overfishing can be, as in the case of the Newfoundland cod collapse, definitively disastrous.

The choices we make about the ocean matter. Sun Myung Moon died in 2012, but his obsession with bluefin tuna and the fish's role in growing the dominance of Unification Church–affiliated fish companies persists. I also still see the echoes of his obsession with the fish resonating in how we collectively regard and consume bluefin. Companies under the umbrella of one of the church's largest affiliated companies, True World Group, continue to build fleets of boats and enormous distribution centers specializing in fresh and frozen seafood, and still hold a huge market share of raw fish distribution in both Canada and the United States. And while one could argue that, given long-standing demand for bluefin in Japan, the global price for tuna would have inevitably risen without the hand of the Unification Church, that's just not how it happened.

As I discovered, as you have discovered alongside me, the future of Atlantic bluefin tuna has hinged on a series of butterfly-wing events: a single black-and-white newspaper clipping tacked up in a Nova Scotia gas station; an idealistic Audubon employee trying to help a fish he loved to catch; a religious leader hearing about a big fish at a tackle shop; a damaged young boy whose uncle once gave him a fishing rod. Those moments all mattered, and those moments are still being made.

EPILOGUE

A BEGINNING

~~~~~~~~~

In December 2019, two months after Lutcavage gave up her Gloucester lab, I was still sitting on the sidewalk as my husband and son ate pizza inside that Upper West Side café in New York. Rapt, I listened as Lutcavage excitedly talked about how the recapture of the fish both she and Al Anderson had tagged more than a decade earlier had been for her "like Christmas." The most exciting thing about the fish, said Lutcavage, was that Amelia had been caught leaving the Mediterranean—where, her group had concluded, the fish had just spawned.

Hanging up from my call with Lutcavage, I felt a frisson through my stomach and chest. For years, I had written about global fisheries and the ocean, dabbling on the outskirts of my fascination with bluefin tuna, drawn by the glamour, pathos, and formless sense of society's obligation to the fish that—unknown to me then—Carl Safina had already worked so hard to instill in people like me for decades. Getting off the phone with Lutcavage, I had the uncanny sense that the pieces of a story were falling into place.

That evening I spent a late night in the musty downstairs computer lab at Columbia University's Pulitzer Hall researching a Rhode Island fisherman I had never heard of before, a man named Captain Al Anderson. His name too common, I added "tuna" to my initial search term to find him, my quick-typing hands glowing in the blue light of my laptop screen. A few more clicks of the mouse and suddenly there he was, dead in this world but still in it, smiling widely out of a memorial announcing his passing. He wore a ball cap; he held a fish proudly; he looked slightly like my father, back when my father had been healthy. A profound sense of loss washed over me—the most selfish writerly grief there is: I couldn't interview someone who wasn't alive anymore. Yet I felt he would be the missing heart of this story, whatever it was.

For days, I mulled the shapes as they formed. I had a mysterious huge fish, a faded yellow tag, the reputation of a man who died before I could reach him. By that point I had stumbled across the vitriolic 2016 online feud between Lutcavage and Safina. Frustratingly, it also appeared as if I had just missed Al. Less than a year had passed since his death, and his widow, a woman his online obituary named as Daryl Anderson, was still mourning her loss. Despite my reticence, I searched for her phone number online and called her on the telephone, leaving an awkward message on a nameless voicemail I wasn't even sure was correct. I found an email address—apparently, I learned online, she had once managed a marina near Narragansett, the town where the couple had lived—but that email address didn't work. So a few days later, still having received no response to my voicemail, I sat down with a stack of printer paper and a pen and wrote Daryl Anderson a letter. I introduced myself and requested a phone conversation or in-person interview with her to learn more about her husband and his decades of fish tagging. Within weeks, my life had become intertwined with the life of this stranger in ways that felt momentous and largely inexplicable. My husband shrugged; I was prone to my own fits of obsession, of endless online searches and rabbit holes. This one, he reasoned, would pass with all the rest.

When Daryl received my letter in the mail, her first instinct, she later told me, was to dodge my request and probing questions entirely. It hurt too much, not having Al around. Not hearing his astute observations, even his bitter complaints, and especially those silly, sweet things he said and did only when they were alone. But then, she recounted to me over mugs of tea in her cozy and spotless Narragansett living room, she had another thought. It felt a bit like Al was talking to her, that her retelling his story of fish tagging was what he wanted, even though he was gone. Why not? that little voice suggested. Just say yes, it said. Perhaps another angler would hear his story, take his advice, and follow the example he had set since he first punctured the skin of those earliest bass with clumsy copper loops. Fishermen who loved the sea had a choice; they should stop merely catching fish and killing them and instead work to better understand, through tagging, those creatures they purported to love. Plus, Daryl reasoned, I had asked nicely.

Building our relationship felt like weaving a delicate spiderweb, probing on my end and guarded on hers, vibrating with the energy and newness of a discovery. It helped that my interest in Al had been sparked by Amelia, that one fish Al had tagged nearly 15 years prior, and that I had been introduced to his work by a scientist who was excited by its transatlantic recapture an ocean away.

Reading about Al and his reputation online, I had inferred that he had been a tough skipper and often a polarizing, maybe sometimes even loathed figure on the Narragansett docks. After his death, fishing websites and comment sections eulogized the captain, passing around his obituary and writing tributes like "Lucky I got to fish with him once," "Thank you for your conservation," and "Sad—Al was a pioneer of [catch and release] where very few practiced it at the time." Although not everyone had fond memories of the mercurial fisherman: "Positive statements you could make about the Captain," another poster wrote, implying he didn't want to talk ill about the dead. "You always knew he was around. He made [h]is presents [sic] known. If he was fishing near you, some how, some time, he was going to have an effect on your day." Another man dryly seconded the sentiment.

During our first conversation Daryl invited me to Rhode Island, and I made the trip a few weeks later. As I pulled up to Daryl's home—once Al's home—my heart jumped into my throat. It was approximately the one-year anniversary of Al's death, and Daryl had invited a handful of his oldest friends and favorite first mates over so they could speak to me about their colleague and mentor. Al's friend and fellow captain Charlie Donilon, the oldest salt in the lot and still an active shark tagger, sat at the table's head, flanked by Al's former first mates Steve Tombs, Paul Osimo, Matty DiMatteo, and Kevin Robichaud.

That first day I spent in Al's home seemed to last forever. It also sped past as I sat around Daryl's wooden kitchen table with five tough, grown men whose eyes brimmed with tears over their memories of Al. Someone passed me a stuffed teddy bear Al often carried onboard as a joke: he often tucked it in with the manliest fisherman on the boat, a bruiser sleeping off his early morning wake-up belowdecks as the *Prowler* headed back to its salt-stained berth at the Snug Harbor Marina. Sometimes he'd take a picture of that big, wide-mouthed charter client, his arm draped around the fuzzy bear, and print it off as a memento. DiMatteo, a former snow crab salesman, held up one of Al's prized inventions, a modified umbrella rig contraption of rubber tubing and wire that Al had tweaked to resemble a swaying nest of sand lance. On it he had caught thousands of striped bass and juvenile bluefin.

Robichaud, a former marine medic and Iraq veteran who fished alongside Al during his final years, told stories of how Al drove the boat in such a specific way past a small school of bluefin that on some days all the rods would hook a fish at the same time, causing the boat's thicket of rods to twang simultaneously. *Pop. Pop. Pop pop pop* and there were tuna everywhere, and the only problem was how to land them all and tag them and set them free without killing any fish.

I went downstairs with Osimo, who had introduced Al at the IGFA gala when he'd joined that association's Legendary rolls and who now teaches at the Rhode Island School of Design. In the mid-aughts, Osimo had responded to an online ad in which Al was looking for a new mate. (Al, the gathered group laughingly told me, always needed new mates; he worked through them like a demanding jockey did a hard-run horse.) Al was a complicated guy, Osimo told me, but the younger angler considered one particular stretch of 14 consecutive days that the two of them fished bluefin in the mid-2000s as some of the hardest and absolute best days of his life. He was fishing beside a living legend, he knew, a fact that helped on days when Al screamed at him for reasons that were often unclear. Osimo and I sat cross-legged on the carpeted floor of Daryl and Al's basement, opening journal after journal, searching for the day on which the far-ranging juvenile bluefin tuna Molly Lutcavage had told me about had been caught.

Suddenly, it was there. We had found it: fall 2004, the name and phone number of the fishermen who had chartered the *Prowler*—Jason Williams—carefully recorded in heavy, looping penmanship. Al had pressed the blue ballpoint pen so heavily into the thin journal paper I could feel the rut left by the rolling ball under my fingertips as I ran them across the page more than a decade after he had written the entry. I had the strange sense of Al's own body being right there with me, him standing at the prow of the *Prowler*, exhausted and satisfied, recording information for his own use, for posterity, or driven by an uncanny sense that someday, just perhaps, that information would produce meaning beyond itself. And here I was.

Propped on my knees in the carpeted basement, I opened cupboards holding stacks, piles, volumes of tagging journals and notebooks, slide decks, and article clippings. Above, on shelves lining the walls, sat gleaming rows of fish jumping atop tagging trophies punctuated by memorabilia and carefully organized books. Framed photographs of Al and his clients and the tagging milestones he had reached hung on every wall: 5,000 fish tagged; 10,000 fish tagged. The numbers climbed to 50,000, then 60,000 fish, and suddenly, Al looked old in the wooden frames. In pictures he almost always wore his wide-brimmed baseball hat and a mustache, his face growing increasingly jowly with age the way my father's had, his smile still bright like my father's had once been. I flipped though his published books and an unpublished manuscript with jumbled chapters he had been writing when he died; one passage jumped out like a message he had sent to me from the past. *"There are many who understand my love and respect for Block Island and its waters,"* he wrote. *"And although its fisheries resources have diminished, my infatuation with the place and its fish have held strong. I hope your experiences on her waters send you home with the satisfaction they've given me. Fair winds, and good luck."*

After that trip to Rhode Island, I couldn't stop thinking about Al, this fisherman few outside his community had heard of before. I spent hours driving around New England ports, asking strangers at stores if they knew of him, trying to find mentions of him in books about tuna or saltwater fishing, and searching for the scraps of memories of him that remained on fishing forums on-line. "Whole industries, whole causes move forward on the shoulders of those willing to hold a course despite ridicule or outright hostility from colleagues," wrote Zach Harvey, a former mate of Al's. "Despite the very real possibility that few if any of their colleagues will notice, much less understand, appreciate or encourage, their work."

There was a story here, I knew it. And so I started working on what would eventually become this book, first within my graduate program, but also as a side project once I had graduated and was hired as an investigative fellow in a documentary newsroom. It wasn't just a story of one fisherman and a single fish, but a story about the whole world, of contested science and corruption, of the miracles and terrors of globalization and its goddamned children: commodification and the perils of human hubris, in driving a species to its absolute limit even as our warming climate transforms the ocean at a dizzying pace.

I started compulsively finding and buying and collecting and digitizing tuna fishing books and memorabilia. I fell in love with the Pablo Neruda poem "Ode to a Large Tuna in the Market," dazzled by its lush morbidity. I called prominent scientists around the world, bearing this tale like a talisman. I told them about Al Anderson, and the story of the bluefin tuna he tagged, then Molly Lutcavage tagged, which then died the same year that he did. A friend's elderly uncle, a veteran of New England tuna fishing, greeted me in his home and, at the interview's end, gifted me an old ball cap for the American Tuna Association along with a tattered 48-page broadsheet entitled *MAN vs. GIANT TUNA: The International Tuna Cup Matches.* The size of a concert poster and the weight of a university textbook, it was a piece of tourist memorabilia produced by the Nova Scotia government in 1973 to promote the province's doomed international tuna tournament. I hadn't found anything similar in any museum or record, and I carried it around as part of my personal library, protecting it from being squashed by books or accidentally recycled. I traveled to Wedgeport on bumpy roads, driving for hours to watch a scientist chainsaw open rotted tuna heads to extract their otoliths with tweezers. I flew to Japan and snuck into Tokyo's new, not-yet-open-to-the-public Toyosu tuna auction building, my mouth agape at the hundreds of fresh and deep-frozen bluefin arrayed on its concrete floors; that morning I ate a plate of maguro nigiri before the sun rose. I bobbed beside the patch of sea where Amelia the tuna had been killed by a shot to the head, and invited myself to dinner at Carl Safina's house, where I ended up chopping us a kale

Caesar salad garnished with bluefish that Safina had caught and smoked himself months before.

It diagramed in a perfect circle. My own hauntings had drawn me toward an obsession with Al Anderson; his own pushed him to understand the world through a lens of science. In every person I interviewed about Al, I saw flashes of my father, in the high-achieving standards he set for himself and those around him, how the absence of a loving, safe-feeling home can tilt a child's world off-balance and leave him, or her, searching for comfort or satisfaction for the rest of his life. But I also saw my father in Carl Safina, in Molly Lutcavage, in those tired of living in a world where they were expected to either be bullies or be bullied and so instead used the orderly rhythm of logic and hypotheses to smooth the world around them. For the first time in a very long time, I felt close to my father in my mind. Sometimes, in a quiet moment, I'd say, "Thanks, Dad," while looking at the sky.

BEFORE I VISITED Rhode Island on my first trip to meet Daryl Anderson, she had emailed me a short, edited-down version of a memoir Al wrote before he died. I printed it out and held the small sheaf in my hands as my bus to Providence crossed a bridge leaving Manhattan. The light of the rising sun strobed through the bridge's girders as Al spoke to me from the pages. I read about his hard, beautiful early memories of living up in a New Jersey neighborhood dubbed the "Dog Patch" for its stray mutts, about how he fell in love with fish science, and about his eventual rise to tagging notoriety. But it was the postscript Daryl later added that hit me hardest as I struggled to control tears that had mostly failed me since my father had died two years before.

As Al neared 70, his health started to fail, this time in a grinding, sad way that felt inevitable, and Daryl worried about him, even on short pleasure trips. Every month the old captain moved more slowly, got thinner, slept more, but despite his failing health, he continued to keep meticulous records, scrawling in his brown, clothbound notebooks with shaky handwriting. On the evening of January 17, 2018, after Al felt chest pains, Daryl called an ambulance that took him to the hospital. Before he died, he told Daryl he didn't have much time but still wanted to finish his final book. Minutes later he was gone, killed by a heart attack: the same death that befell his father. Al died only a few months before Amelia was recaptured off the Algarve, and he would have delighted in the news of her third and final capture. Perhaps he would have mourned her as well.

Years before, Al had told Daryl that he wanted his ashes scattered at his favorite fishing spot, the North Rip off Block Island. In typical Al fashion, he provided her with the exact GPS coordinates. "At first, I was taken aback by this idea," she wrote in her postscript. She had, after all, grown up with a father who was a commercial fisherman and she knew too many men who had drowned at sea. To cover her discomfort, she cracked a joke that if Al didn't behave himself, she would instead "place his ashes in an hourglass on the hutch and make him keep working." That horrified Al, but she quickly assured him that it was only a joke. Of course, she said gently, she would honor his wishes.

On a foggy morning four days after a full moon, Daryl and a handful of Al's friends and former mates motored past "the gap," or breakwater, on the *Snappa*, a boat captained by Al's longtime friend Charlie Donilon. For most of the trip out, Daryl stayed in the *Snappa*'s cabin, overwhelmed by the experience of traversing for the last time, the same path her beloved skipper had navigated so many thousands of times before.

Once they reached the North Rip, the group gathered in the boat's cockpit and Daryl raised a red tagging flag to half-mast on the boat's outrigger. Donilon, who has legally performed ash burials at sea for decades, led the ceremony as the group shared their favorite memories of Al. They played his treasured rock tunes from the boat's tinny speakers as they cried and laughed.

Reading the end of that letter on the bus ride to Rhode Island—before I'd ever met Daryl, seen a giant bluefin tuna, or connected my own story of loss and love with Al's own—I found myself overcome. *"After a final prayer it was time for the basket with Alan's ashes to be lowered from the stern of the boat,"* Daryl wrote on the final page of Al's last, and forever unpublished, book. *"The seas were glassy and calm; the basket left my hand and settled gently to the bottom releasing a luminous cloud. The boat circled the yellow roses floating in the tide twice and then the ship's bell rang eight times to signal that Alan's watch was over."* The sun refracted through my tears as my bus barreled toward the story that awaited me, and I couldn't wait to begin.

ACKNOWLEDGMENTS

As you've gathered, Daryl Anderson is this book's patron saint. She opened her heart and her home to me during one of the hardest times in her life and granted access to her most precious and worst memories of her 40-year-long marriage. She welcomed me for repeated visits to dig through Al's trove of papers, photographs, videos, and carefully organized paraphernalia. Daryl, I am so grateful and have done my very best to do your lives together justice. Also in New England, thanks to Al's family, fishing colleagues, friends, and nemeses who helped me reconstruct decades of his life, including Susanne Devine, Matty DiMatteo, Charlie Donilon, Janet Malenfant, Paul Osimo, Al Ristori, Kevin Robichaud, Bill Sisson, and Steve Tombs.

To write this book, I drew on the expertise and generosity of time of a wide array of brilliant scientists, researchers, and anglers. Particular thanks to Scott Barrett, Andre Boustany, Doug Butterworth, Matthew Corporon, Paul DeBruyn, Jack Devnew, Jean-Marc Fromentin, Jaime Gibbon, Zach Harvey, Christina Hernández, Alan Hokanson, Paul Howey, Camille and Eric Jacquard, Scott Kraus, Susan and Roger Lema, Pedro Lino, Molly

Lutcavage, Anthony Mendillo, Shana Miller, Arturo Morales-Muñiz, J. J. Maguire, Daniel Pauly, Alfredo Poço, Gregory Puncher, Andrew Rosenberg, Rich Ruais, Carl Safina, Laurenne Schiller, Mike Sissenwine, Jason Williams, Boris Worm, and all those who asked to remain unnamed or anonymous. You might not always agree with one another, but I believe that history is the ultimate arbiter and that you're all doing your best in your respective ways.

I'm particularly grateful to Shanna Baker, Jesse Hirsch, Abby Johnston, Rafil Kroll-Zaidi, Harley Rustad, and Jan Wong for opportunities, advice, and edits along the way. At Columbia University, professors Marguerite Holloway, Jonathan Weiner, and Paige West helped me weave a new intellectual cocoon; the inimitable Sam Freedman touched a spark to the fuse. I owe much to Keith Goggin, my first and best patron, and his alumni class of 1991, as well as to other friends and colleagues of Pulitzer Hall, especially my dear Freedmanites.

Many thoughtful books have been written on global fisheries and bluefin tuna, and it was a delight to devour as many as I could. I remain particularly in awe of work by Stephen Adolf, Carmel Finley, Paul Greenberg, Mark Kurlansky, Theresa Maggio, Carl Safina, Jennifer Telesca, Laura Trethewey, and Douglas Whynott. In Spain and Portugal, my reporting wouldn't have been possible without dogged fixing and translating work from journalist Ulises Izquierdo. (*Ei!* We did it!) I'm also grateful to Annie Manson and Jane Cooney in Cádiz, and Daisuke Hayashi in Japan.

My extraordinary agent, Mackenzie Brady Watson, at Stuart Krichevsky Literary Agency, believed in this book from the moment she yanked it from her slush pile and has since become a treasured friend and confidant. Lynn Henry and Pamela Murray at Knopf Canada and Myles Archibald at William Collins made early bets on me and Al; John Parsley and Cassidy Sachs at Dutton committed to this book with their huge hearts and brains, and this book you now hold is a million times better as a result. I worship Lorie Pagnozzi's book design. I am grateful to every brilliant person on my publishing teams for their heroic efforts.

This book wouldn't have been possible without grants from the Canada Council for the Arts and the Sloan Foundation. Their support of this project allowed me to pay for fact-checking from the outrageously talented Rhiannon Russell, who saved me from myself repeatedly. I finished writing this book (and then burned my error-riddled manuscript) at the Logan Nonfiction Program and found solace and inspiration from the incredible journalists and creators there. Thanks especially to Alice Driver and Max Duncan: Swim Club forever.

My friends and chosen family around the world and online are woven throughout this book and the ideas explored within it; your brilliance (and brilliant fixes and edits!) influenced my thinking and rethinking throughout. Special thanks to Hildy, Matt and Chelsea, Lizzie, Grace "Panther" Williams, Simon, Lyndsie, Erin, Corey "Peacemaker" McLean, Eric, Jane, Kira, Susan, and Glenn.

My mother, Lydia Luckevich, provided my earliest example of how to carve a place for yourself in a man's world and helped make this book a reality. (She also saved both me and you, dear reader, from the phrase "milty sperm.") My siblings, Veronica and Graham, and their partners, Clay and Diana, cheered me on from the sidelines. My in-laws, Pat and Rick, provided childcare and boundless enthusiasm when mine threatened to wane. In my early life, my dziadzia—the incredible Alexander Luckevich—passed on a passion for current affairs that often buoyed my spirits when a career in journalism felt unhinged. My beloved father, Don Pinchin, brought me so far into this world and left it so much poorer when he passed. This one's for you, Dad.

My partner in crime and incredible husband, Sean Sullivan, put his life on hold for two years while I chased tuna dreams around the world, only to find ourselves back where we started. This wouldn't have been possible without your love, boundless support, and top-notch edits, SPS. And finally, to my precious son, Parker James: I love you more than the moon and the stars. I'm trying to leave this world better than I found it because of you.

NOTES

CHAPTER ONE: HOOKED

1. That began to change in South Africa in the 1960s, as animal rights groups there accused anglers of animal cruelty, but that debate didn't spill into the United States until the mid-1990s. (Troy Vettese, Becca Franks, and Jennifer Jacquet, "The Great Fish Pain Debate," *Issues in Science and Technology* 36, no. 4 (summer 2020): 49–53.)

2. While purse seining on an industrial scale rose dramatically after the Second World War, the first recorded purse seine fishery for bluefin started in 1929 in Yugoslavia. (Whynott, p. 79)

CHAPTER TWO: "YOU DAMNED WELL BETTER TAG"

1. The origin of the design, first claimed by British inventor Sir Charles Dennistoun Burney, has since been contested by a retired Icelandic marine engineer, who claims Burney stole it from Icelander Andres Gunnarsson after the latter made a failed trip to Britain to sell the idea. (Keith Findlay, "New Claim on Founding of Stern-Trawler Design," *The Press and Journal*, August 3, 2018.)

CHAPTER FOUR: BEFORE THE STORM

1. NOAA eventually published Mather's report in July 1995, 20 years after it was first drafted.

2. This later became the Magnuson-Stevens Fishery Conservation and Management Act when it was amended in 1996.

3. The Atlantic Tuna Club has been based out of Block Island since 1915 and is the oldest tuna-fishing club in the Northeast.

CHAPTER FIVE: RISING MOON, FLYING FISH

1. In 1975, Cyganowski joined Roger Hillhouse in adhering to the 1,100-ton self-imposed quota. (Whynott, p. 84)

CHAPTER SIX: RED GOLD

1. In the mid-2000s, as documented by environmental governance professor Jennifer E. Telesca in her book *Red Gold*, a Croatian delegate to ICCAT once declared, "We love our tuna more than our women," as he waved one of these coins during one international meeting. (Telesca, p. 62)

2. Once, about eight years ago, I made my own homemade garum out of a slurry of smelt guts, coarse salt, and pineapple juice, which provides the enzymes needed for bacteria-free fermentation. I put it in a plastic container, poked holes in the top, and let it sit at room temperature for months. It didn't stink as much as I expected, and made a glorious-tasting condiment for soups and red meats like lamb.

CHAPTER SEVEN: KINGS OF THE OCEAN

1. Calling a population of fish "stocks" is a term modern environmentalists have long condemned for its reduction of living animals into items for trade. I tend to agree with the flattening effect the term can have on valuing the life of marine creatures and try to avoid the term.

2. Previously, in 1978, the Canadian delegation had put the idea of splitting the stock forward to a vote at ICCAT, but it was never voted on. (Devnew, p. 75)

CHAPTER NINE: AMELIA

1. Now called the New England Fishery Management Council.

CHAPTER TEN: "THEY WILL COME FROM OUTSIDE"

1. The forerunner to Operation Tarantelo was Operation Opson VII, a 2017–18 investigation by European police into the illegal tuna market. At its culmination, 45 tonnes of tuna were seized.

SELECTED BIBLIOGRAPHY

EPIGRAPH

Neruda, Pablo. "Oda a un gran atún en el mercado." *Tercer libro de las odas.* © Pablo Neruda, 1957, and Fundación Pablo Neruda. Translated by Robin Robertson. *Poetry,* April 2007.

PROLOGUE

Whynott, Douglas. *Giant Bluefin.* North Point Press, 1996.

CHAPTER ONE: HOOKED

Anderson, Al. *The Atlantic Bluefin Tuna: Yesterday, Today, and Tomorrow.* Fisherman Library, 1990.

———. *Island Stripers: A Fisherman's Guide to Block Island Bass.* Xlibris US, 2012.

———. *60,000 Game Fish.* Unpublished manuscript.

———. *To Catch a Tuna.* MT Publications, 1990.

Barry, Dave. "Blub Story." *Tropic* (*Miami Herald* Sunday magazine), February 19, 1989.

Carey, Francis G., and John M. Teal. "Heat Conservation in Tuna Fish Muscle." *PNAS* 56, no. 5 (November 1966).

Ellis, Richard. *Tuna: A Love Story.* Knopf, 2008.

Magnusson, Harris W. "Freezing Fish at Sea—New England. Part 10— Studies of Miscellaneous Handling Problems." *Commercial Fisheries Review* 17, no. 7 (July 1955).

Muhling, Barbara, et al. "Predicting the Effects of Climate Change on Bluefin Tuna (*Thunnus thynnus*) Spawning Habitat in the Gulf of Mexico." *ICES Journal of Marine Science* 68, no. 6 (2011): 1051–62.

Nakamura, Y., K. Mori, K. Saitoh, et al. "Evolutionary Changes of Multiple Visual Pigment Genes in the Complete Genome of Pacific Bluefin Tuna." *PNAS* 110, no. 27 (2013): 11061–66.

O'Connor, Constance M. "The Physiological Consequences of Catch-and-Release Angling: Perspectives on Experimental Design, Interpretation, Extrapolation and Relevance to Stakeholders." *Fisheries Management and Ecology* 20 (2013): 268–87.

Planet Tuna. "The Life Cycle of the Atlantic Bluefin Tuna." March 18, 2019. https://planettuna.com/en/the-life-cycle-of-the-atlantic-bluefin-tuna-how-a-3-mm-larva-turns-into-a-400-kg-giant/.

Safina, Carl. *Song for the Blue Ocean: Encounters Along the World's Coasts and Beneath the Seas.* Henry Holt, 1998.

CHAPTER TWO: "YOU DAMNED WELL BETTER TAG"

Aguilera, Mario. "Ocean Noise Has Increased Considerably Since 1960s, According to New Scripps Analysis." UC San Diego news release, August 18, 2006.

Alexander, Lewis M., and R. D. Hodgson. "The Impact of the 200-Mile Economic Zone on the Law of the Sea." *San Diego Law Review* 12 (1975): 569.

Bryant, Nelson. "Wood, Field and Stream: Fish Tagging." *New York Times*, April 27, 1975.

Carson, Rachel. "Undersea." *The Atlantic*, September 1937.

Corbyn, Zöe. "Archaeologists Land World's Oldest Fish Hook." *Nature* (2011).

Finley, Carmel, and Naomi Oreskes. "Maximum Sustained Yield: A Policy Disguised as Science." *ICES Journal of Marine Science* 70, no. 2 (March 2013): 245–50.

Lackey, Robert T. "Fisheries: History, Science, and Management." In *Water Encyclopedia: Surface and Agricultural Water*, edited by Jay H. Lehr and Jack Keeley. John Wiley and Sons, 2005.

Larkin, P. A. "An Epitaph for the Concept of Maximum Sustained Yield." *Transactions of the American Fisheries Society* 106, no. 1 (1977): 1–11.

Lutcavage, Molly. "Bluefin Spawning in Central North Atlantic?" *Pelagic Fisheries Research Program* 6, no. 2 (April–June 2001).

Lutcavage, Molly, and Scott Kraus. "The Feasibility of Direct Photographic Assessment of Giant Bluefin Tuna, *Thunnus thynnus*, in New England Waters." *Fishery Bulletin* 93, no. 3 (1995).

Royce, William F. "Historical Development of Fisheries Science and Management." Lecture, Fisheries Centennial Celebration, 1985.

Sisson, William. "Wandering Marvels." *Anglers Journal*, June 8, 2018.

Wang, Junshi, D. Wainwright, R. Lindengren, et al. "Tuna Locomotion: A Computational Hydrodynamic Analysis of Finlet Function." *Journal of the Royal Society Interface* 17 (April 2020).

CHAPTER THREE: AGE OF GIANTS

Arundel, Wendy. "The Strange Tale of Outer Baldonia: A Fisherman and His Island Micro-nation." *Maine Boats, Homes, and Harbors* 163 (March–April 2020).

Coe, Nancy C. "Outer Baldonia Struggles for Its Fishy Place as a Somewhat Emergent Nation." *Sports Illustrated*, September 18, 1967.

Virtual Museum Canada. "Wedgeport Tuna Tournaments Past and Present." Community Stories Collection. Wedgeport Sport Tuna Fishing Museum and Interpretive Centre, 2012.

Fisheries and Oceans Canada, Communications Branch. *Pelagic Fish: Marathon Swimmers: Scotia-Fundy Region*. March 1994.

Gifford, Tommy. *Anglers and Muscleheads*. E. P. Dutton, 1960.

Heilner, Van Campen. *Salt Water Fishing*. Penn, 1937.

Jacquard, Donnie. "Amelia Earhart and Wedgeport Tuna Fishermen." *The Argus* 21, no. 1 (Spring 2009).

MacDonald, David. "The Maddest Three Days in Fishing." *Maclean's*, September 17, 1955. https://www.biblio.com/book/macleans-magazine -september-17-1955-rocket/d/491961613.

Mackinnon, Lachlan. "The Principality of Outer Baldonia: A Nova Scotia Micronation." *Active History*, November 27, 2014.

Perry, Margaret. *Battling "Blue-Fins."* National Film Board of Canada, 1947.

Varga, Darrell. "Margaret Perry and the Nova Scotia Film Bureau." *The Way We Were: Nova Scotia in Film, 1917–1950.* Nova Scotia Archives, 2010.

CHAPTER FOUR: BEFORE THE STORM

Almy, Gerald. "Tuna Catch Gives Hope for Future." *Washington Post,* July 6, 1979.

Cole, John N. "The Vanishing Tuna." *The Atlantic,* December 1976.

Kennan, George F. "To Prevent a World Wasteland." *Foreign Affairs* 48, no. 3 (April 1970).

Mather, Frank J., III, John M. Mason Jr., and Albert C. Jones. *Life History and Fisheries of Atlantic Bluefin Tuna.* US Department of Commerce, National Oceanic and Atmospheric Administration, National Marine Fisheries Service, Southeast Fisheries Science Center, 1995.

Murphy, John F. *Newport, Block Island and Narragansett Pier.* John F. Murphy, 1903.

Nebiker, Walter, Robert Owen Jones, and Charlene K. Roice. *Historic and Architectural Resources of Narragansett, Rhode Island.* Rhode Island Historical Preservation Commission, 1991.

Nyqvist, Daniel, et al. "Electric and Magnetic Senses in Marine Animals, and Potential Behavioral Effects of Electromagnetic Surveys." *Marine Environmental Research* 155 (2020).

Reiger, George. "Frank Mather—in Memoriam." *Field and Stream,* August 2000. https://books.google.ca/books?id=HkzeZVg62uQC&lpg=PA28&ots =V3aXeXJZac&dq=frank%20mather%20field%20and%20stream&pg=PA28 #v=onepage&q=frank%20mather%20field%20and%20stream&f=false.

Ruais, Rich. "Requiem for 'Iconic Bluefin' Extinction Myth: A Seven-Part Series." *Commercial Fisheries News,* August 2011–February 2012.

Wilde, Oscar. "The Critic as Artist." In *Intentions.* Osgood, McIlvaine, 1891.

CHAPTER FIVE: RISING MOON, FLYING FISH

Bland, Alastair. "From Cat Food to Sushi Counter: The Strange Rise of the Bluefin Tuna." *Smithsonian Magazine,* September 11, 2013.

Buchanan, Daniel Crump. *Japanese Proverbs and Sayings.* University of Oklahoma Press, 1965.

Carey, Francis G. "Fishes with Warm Bodies." *Scientific American* 228, no. 2 (February 1973): 36–45.

Corson, Trevor. *The Story of Sushi: An Unlikely Saga of Raw Fish and Rice.* Harper Perennial, 2008.

———. "Why I Don't Miss Bluefin Sushi." *The Atlantic,* November 19, 2009.

Daniel, Leon. "'Moonies' Controversy Flares in Famed Fishing Town." United Press International, May 14, 1981.

Demilo, David A. "God's Catch." *Harvard Crimson,* September 19, 1979.

Franklin, Ben A. "Moon Church's Flotilla Strikes a Reef." *New York Times,* July 6, 1984.

Fromson, Daniel. "The Untold Story of Sushi in America." *New York Times,* November 5, 2021.

Horowitz, Sara. "Vancouver's Legendary Sushi Chef Hidekazu Tojo Invented Far More Than the California Roll." *Montecristo Magazine,* December 11, 2017.

Hokanson, Allan. *Fish Follow the Fisherman.* Lulu, 2016.

House, Jonas. "Sushi in the United States, 1945–1970." *Food and Foodways* 26, no. 1 (2018): 40–62.

Hyer, Marjorie. "Massachusetts Town Mobilizes Against Moonies." *Washington Post,* August 25, 1980.

Inglis, Michael., ed., Peemoeller, Gerhard in *40 Years in America.* I Publications, 2000. https://www.tparents.org/Library/Unification/Books/40Years/40-4-43.htm

Issenberg, Sasha. "The One That Almost Got Away." *Boston* magazine, April 25, 2007.

———. *The Sushi Economy.* Penguin Random House Canada, 2008.

Itoh, Makiko. "Once Considered Low Class, How Did Tuna Get So Valuable?" *Japan Times,* June 17, 2016.

Moon, Sun Myung. "The Way of Tuna." Speech in Belvedere, New York, July 13, 1980. Translated by Bo Hi Pak. http://www.tparents.org/Moon-Talks/sunmyungmoon80/SunMyungMoon-800713a.htm.

Moreno, Edward. "Sushi in Los Angeles: Part 1." *Discover Nikkei,* August 3, 2010.

Mower, Joan. "Moonie Tuna Tournament Draws Criticism." United Press International, August 15, 1981.

Parkinson, David. "When Fish Began to Fly." *Globe and Mail*, July 12, 2007.

Porch, Clay E., et al. "The Journey from Overfishing to Sustainability for Atlantic Bluefin Tuna, *Thunnus thynnus.*" In *The Future of Bluefin Tunas: Ecology, Fisheries Management, and Conservation*, edited by Barbara A. Block. Johns Hopkins University Press, 2019.

Rosenthal, Jack. "The Editorial Notebook; Sushi at the Harvard Club." *New York Times*, November 2, 1981.

SanDiegoYuYu. "YuYu Interview Noritoshi Kanai," April 30, 2007.

Smith, Andrew F. *American Tuna: The Rise and Fall of an Improbable Food*. University of California Press, 2012.

Sullivan, Kevin, and Mary Jordan. "Once-Generous Japanese Become Disenchanted with Moon's Church." *Washington Post*, August 4, 1996.

The Sushi Geek. "Hanaya Yohei and the Beginning of Nigiri-Zushi." TheSushiGeek.com, January 24, 2016.

Thompson, Reg. "Clow, Roy T11." Island Voices oral history. University of Prince Edward Island, 2006.

White, Madeleine. "Meet the Man Behind the California Roll." *Globe and Mail*, October 23, 2012.

CHAPTER SIX: RED GOLD

Adolf, Steven. *Tuna Wars: Powers Around the Fish We Love to Conserve*. Springer, 2019.

Bernal-Casasola, Dario. "Whale Hunting in the Strait of Gibraltar During the Roman Period?" *SAA Archaeological Record* 18, no. 4 (September 2018).

Calkins, Mandy. "Following the Light." *Science and the Sea*, September 1, 2011.

Camiñas, Juan Antonio, et al. *Killer Whale*, Orcinus orca *(Linnaeus, 1758) in the Strait of Gibraltar and Interactions with Spanish Tuna Fisheries*. Final report and addendum to the MoU ACCOBAMS No. 06/2016/LB 6410, July 2018.

Cort, José Luis, and Pablo Abaunza. "The Bluefin Tuna Catch in the Strait of Gibraltar: A Review of Its History." In *The Bluefin Tuna Fishery in the Bay of Biscay*. SpringerBriefs in Biology, 2019.

————. "The Fall of the Tuna Traps and the Collapse of the Atlantic Bluefin Tuna, *Thunnus thynnus* (L.), Fisheries of Northern Europe from the 1960s." *Reviews in Fisheries Science and Aquaculture* 23, no. 4 (September 2015): 346–73.

Di Natale, Antonio. "The Ancient Distribution of Bluefin Tuna Fishery: How Coins Can Improve Our Knowledge." *Collective Volume of Scientific Papers ICCAT* 70, no. 6 (January 2014): 2828–44.

Eyüboğlu, Bedri Rahmi. "The Saga of Istanbul." *A Brave New Quest: 100 Modern Turkish Poems*, edited and translated by Talat S. Halman, associate editor Jayne L. Warner, 33–39. Syracuse University Press, 2006. © Syracuse University Press. Reproduced with permission from the publisher.

Longo, Stefano B., Rebecca Clausen, and Brett Clark. "From Tuna Traps to Ranches." Chapter 4 in *Tragedy of the Commodity: Oceans, Fisheries, and Aquaculture*. Rutgers University Press, 2015.

Maggio, Theresa. *Mattanza: Love and Death in the Sea of Sicily*. Perseus Books Group, 2000.

McPhee, John. *The Founding Fish*. Farrar, Straus and Giroux, 2003.

Morales-Muñiz, Arturo. "Where Are the Tunas? Ancient Iberian Fishing Industries from an Archaeozoological Perspective." *Skeletons in Her Cupboard: Festschrift for Juliet Clutton-Brock*. Oxbow Monographs, vol. 34, 1993.

Puncher, Gregory N., Alex Hanke, Dheeraj Busawon, et al. "Individual Assignment of Atlantic Bluefin Tuna in the Northwestern Atlantic Ocean Using Single Nucleotide Polymorphisms Reveals an Increasing Proportion of Migrants from the Eastern Atlantic Ocean." *Canadian Journal of Fisheries and Aquatic Sciences* 79, no. 1 (June 2021).

Puncher, Gregory N., Vedat Onar, Nezir Yasar Toker, and Fausto Tini. "A Multitude of Byzantine Era Bluefin Tuna and Swordfish Bones Uncovered in Istanbul, Turkey." *Collective Volume of Scientific Papers ICCAT* 71, no. 4 (2015): 1626–31.

Telesca, Jennifer E. *Red Gold: The Managed Extinction of the Giant Bluefin Tuna*. University of Minnesota Press, 2020.

CHAPTER SEVEN: KINGS OF THE OCEAN

Butler, Michael J. A. "Plight of the Bluefin Tuna." *National Geographic*, 1982.

Devine, Denis. "How U.S. Nonprofits Came to Care About Fish." *Driftwords*, 2008.

Devnew, Jack. "Politics and the Conservation of Atlantic Bluefin Tuna." MA thesis, University of Delaware, Department of Marine Policy, 1983.

Fogt, Jan. "For Better or Worse?" *Boating*, January 1988.

Garland, Susan. "Bluefin Tuna: Saved from Nets After Scientists Find Severe Shortages." *Christian Science Monitor*, December 8, 1981.

Gordon, William G. *End of the Year Report for 1982*. NOAA, February 14, 1983.

Havice, Elizabeth, et al. "New Data Technologies and the Politics of Scale in Environmental Management." *Annals of the American Association of Geographers* 112, no. 8 (2022).

Korman, Seth. "International Management of the Atlantic Bluefin: Political and Property-Rights Solutions." *Virginia Journal of International Law* 51, no. 3 (June 7, 2010).

Levin, Dan. "End of the Salad Days for Bluefin Tuna." *Sports Illustrated*, November 18, 1974.

Seabrook, John. "Death of a Giant." *Harper's Magazine*, June 1994.

Sisson, William. "Wandering Marvels." *Anglers Journal*, June 8, 2018.

Sonu, Sunee C. *Japan's Tuna Market*. NOAA Technical Memorandum NMFS, September 1991.

Vitalis, Daniel. "The Truth About Bluefin Tuna with Molly Lutcavage." *WildFed Podcast*, episode 16, February 13, 2020.

Williamson, G. R. *The Bluefin Tuna in Newfoundland Waters*. Newfoundland Tourist Development Office, 1962.

US Congress. *Atlantic Tuna: Hearings Before the Subcommittee on Fisheries and Wildlife Conservation and the Environment and the Committee on Merchant Marine and Fisheries, House of Representatives, Ninety-sixth Congress, second session, on Atlantic Tuna Convention authorization and oversight, H.R. 6310. March 6, 1980*. Government Printing Office, 1980.

CHAPTER EIGHT: THE EXTINCTION AGENDA

Block, Barbara. "Hot Tuna: Electronic Tracking of Giant Bluefin Across the Open Sea." AAAS annual meeting in Seattle, Washington, 2004 (reformatted by Stanford University Libraries, 2017).

Crockett, Lee. "The Fish That Inspired a Woman to Help Save a Species." *PEW,* October 30, 2014.

Ellis, Richard. *Tuna: Love, Death and Mercury.* Knopf Doubleday, 2009.

Harvey, Zach. "A Tuna, an Old Fish Tag and the Captain Behind It All." *Soundings,* March 30, 2014.

Stanford University. *Bluefin Tuna Tagging, B-Roll,* videotape. News and Publications Service, August 16, 2001.

US House of Representatives. *Conservation and Management of Highly Migratory Species: Hearing Before the Subcommittee on Fisheries and Wildlife Conservation and the Environment of the Committee on Merchant Marine and Fisheries, 102nd Congress, second session, May 27, 1992.* Government Printing Office, 1992.

CHAPTER NINE: AMELIA

Baxter, Anna. "CITES Fails to Protect Atlantic Bluefin Tuna." Oceana press release, March 18, 2010.

Census of Marine Life. "Collapse of Bluefin Tuna Population off Northern Europe Described." ScienceDaily, August 7, 2007.

Galuardi, Benjamin, and Molly Lutcavage. "Dispersal Routes and Habitat Utilization of Juvenile Atlantic Bluefin Tuna, *Thunnus thynnus,* Tracked with Mini PSAT and Archival Tags." *PLoS ONE,* May 22, 2012.

International Game Fish Association. 2015 IGFA Tommy Gifford Awards— Legendary Captains and Crew ceremony, posted online May 7, 2020.

Muir, John. *My First Summer in the Sierra.* Houghton Mifflin, 1911; reprinted by Sierra Club Books, 1988.

CHAPTER TEN: "THEY WILL COME FROM OUTSIDE"

Baker, Billy. "In Her Garage Lab, a Scientist Looks for Answers About Skinny Tuna." *Boston Globe,* September 26, 2019.

Center for Biological Diversity. "Illegal Fishing Threatens to Drive Eastern Atlantic Bluefin Tuna Extinct." Press release, October 18, 2011.

Europol. "How the Illegal Bluefin Tuna Market Made over EUR 12 Million a Year Selling Fish in Spain." Press release, October 16, 2018.

Garcia Vargas, Enrique, and David Florido del Corral. "The Origin and Development of Tuna Fishing Nets (*Almadrabas*)." *Ancient Nets and Fishing Gear: Proceedings of the International Workshop on "Nets and Fishing Gear in Classical Antiquity: A First Approach."* Universidad de Cadiz, Servico de Publicaciones and Aarhus University Press, 2007.

International Commission for the Conservation of Atlantic Tunas. *Report for Biennial Period, 2004–05, Part I (2004)—Vol. 1,* English version. ICCAT, 2005.

Whitworth, Joe. "45 Tons of Frozen Tuna Seized in Spain Because of Illegal Treatment." *Food Safety News,* August 16, 2018.

Workshop on the Future of Atlantic Bluefin Tuna: Fisheries, Management and the Market. Final report, proceedings February 3–4, 2021.

CHAPTER ELEVEN: BEGINNING'S END

Arrizabalaga, Haritz, et al. "Life History and Migrations of Mediterranean Bluefin Tuna." *The Future of Bluefin Tunas: Ecology, Fisheries Management, and Conservation.* JHU Press, August 6, 2019.

Carruthers, Tom. "So, You're Considering a Multi-Stock, Spatial, Seasonal Model Eh?" University of British Columbia IOF Quantitative Seminar Series, December 1, 2020.

Dean, Cornelia. "Study Sees 'Global Collapse' of Fish Species." *New York Times,* November 3, 2006.

Di Natale, Antonio. "Due to the New Scientific Knowledge, Is It Time to Reconsider the Stock Composition of the Atlantic Bluefin Tuna?" *Collective Volume of Scientific Papers ICCAT* 75, no. 6 (2019): 1282–92.

Gordoa, Ana. "Determination of Temporal Spawning Patterns and Hatching Time in Response to Temperature of Atlantic Bluefin Tuna (*Thunnus thynnus*) in the Western Mediterranean." *PLoS ONE,* March 7, 2014.

Greenpeace. "Paris: Activists Dump 5 Tons of Bluefin Tuna Heads." Press release, November 18, 2008.

Reglero, Patricia, et al. "Atlantic Bluefin Tuna Spawn at Suboptimal Temperatures for Their Offspring." *Proceedings of the Royal Society Biological Sciences,* January 10, 2018.

INTERVIEWS CITED

Daryl Anderson, Scott Barrett, Marty Bartlett, Andre Boustany, Dheeraj Busawon, Doug Butterworth, Matthew Corporon, Paul DeBruyn, Susanne Devine, Jack Devnew, Matty DiMatteo, Charlie Donilon, Jaime Gibbon, Zach Harvey, Isaac Hermo Cespon, Marcos García Rey, Paul Greenberg, Rafael Márquez Guzmán, Christina Hernández, Alan Hokanson, Paul Howey, Camille and Eric Jacquard, Jean-Marc Fromentin, Scott Kraus, Sylvie Lapointe, Susan and Roger Lema, Pedro Lino, Molly Lutcavage, Annie Manson, Anthony Mendillo Jr., Shana Miller, Matthew Monks, Arturo Morales-Muñiz, J. J. Maguire, Bernard G. Mihalko, Paul Osimo, Mario Payán, Daniel Pauly, Alfredo Poço, Gregory Puncher, Al Ristori, Paloma de la Riva, Kevin Robichaud, Andrew Rosenberg, Rich Ruais, Carl Safina, Laurenne Schiller, Mike Sissenwine, Bill Sisson, Ignacio Soto Medina, Mike Stokesbury, Jennifer Telesca, Steve Tombs, Jason Williams, and Boris Worm.

INDEX

Abrams, Gerry, 166
Adolf, Steven, 135, 148, 223
Albatross III, 26
Albert II (Prince of Monaco), 222–223
almadraba, 141, 142, 143, 144, 148,
    229–232, 239–240, 243
Alper, Leo, 120
Amelia
    Anderson's tagging of, 218
    data points from recapture of, 256
    hatching of, 16
    as juvenile, 16–17
    Lutcavage and, 219–222
    recapture and death of, 250
    size of, 218, 220
    spawning by, 249–250
    tag number of, 255
    two-stock theory and, 10–11
American Littoral Society (ALS), 34,
    98–99
Anderson, Al
    basic facts about, 17, 30, 31, 33–34,
        45, 53, 94, 95–96, 182–183,
        272, 274
    binge drinking by, 191–192
    as biology teacher, 83, 96–97,
        155–156
    books by, 170, 178, 213–214, 215, 226
    brain tumor of, 185–186

charter business and classes of, 55,
    97–98, 101–102, 103, 155–156,
    179–181, 182, 183–185, 187, 188,
    193–194, 215, 217–219, 275
childhood of, 23–25, 27–29, 95
Daryl and, 91–92, 152–155
death and burial of ashes of,
    280–281
marriages, 31, 93–94, 193
on odor of dead fish in cars, 38
stories told about, 274–276
Tombs and, 168–169
Anderson, Al and tagging
    awards for, 10
    beginning of, 32–33
    as centerpiece of charters, 194
    charter business and, 55, 97–98,
        101–102, 103, 155–156, 179–181,
        182, 183–185, 187, 188, 193–194,
        215, 217–219, 275
    first recapture by someone else of
        fish tagged by Al, 98–99, 158
    first recapture of fish previously
        tagged by self, 21–23
    as foundational, 262
    importance of, 181
    Mather and, 44, 46
    method used, 45–46
    plastic darts and, 44

INDEX